城市行走书系
策划：江岱，姜庆共

上海武康路建筑地图
文字：乔争月
摄影：张雪飞，邵律，乔争月
绘图：孙晓悦

责任编辑：徐希
助理编辑：周希冉
书籍设计：孙晓悦
英文校对：Stehpen P. Davis

鸣谢：
郑时龄，章明，伍江，常青，钱宗灏，
宋浩杰，姜江，吴志伟，沙永杰，周立民，
王林，华霞虹，刘刚，曹永康，钱锋，陆烨，
卢卡·彭切里尼，江似虹，卡罗林·罗伯逊，
陈乔丹，陈贤

上海市档案馆
上海市图书馆
上海市人民政府新闻办公室
上海徐家汇藏书楼
上海日报社
新民晚报社
上海市房地产科学研究院
徐汇区规划和土地管理局
徐汇区住房保障和房屋管理局
徐汇区文化局
徐汇区旅游局
徐汇区旅游公共服务中心
湖南路街道办事处
徐房集团
上海宋庆龄故居纪念馆
上海巴金故居
上海章明建筑设计事务所
上海汽车工业总公司
南京墨辰影像传媒有限公司

同济大学出版社
Tongji University Press

CityWalk Series
Curator: Jiang Dai, Jiang Qinggong

Shanghai Wukang Road
Text: Michelle Qiao
Photograph: Zhang Xuefei, Shao Lv, Michelle Qiao
Illustration: Sun Xiaoyue

Editor: Xu Xi
Assistant Editor: Zhou Xiran
Book Designer: Sun Xiaoyue
Proofreading: Stehpen P. Davis

Acknowledgements:
Zheng Shiling, Zhang Ming, Wu Jiang,
Chang Qing, Qian Zonghao, Song Haojie,
Jiang Jiang, Wu Zhiwei, Sha Yongjie, Zhou Limin,
Wang Lin, Hua Xiahong, Liu Gang, Cao Yongkang,
Qian Feng, Lu Ye, Luca Poncellini, Tess Johnston,
Carolyn Robertson, Chen Qiaodan, Chen Xian

Shanghai Municipal Archives
Shanghai Library
Information Office of Shanghai Municipality
The Xujiahui Library
Shanghai Daily
Xinmin Evening News
Shanghai Real Estate Science Research Institute
Xuhui District Town Planning Administrative Bureau
Xuhui District Real Estate Administrative Bureau
Xuhui District Culture Bureau
Xuhui District Tourism Bureau
Xuhui Historical Tourism Service Center
Hunan Road Subdistrict
Shanghai Xufang Group
Shanghai Song Chingling Memorial Residence
Ba Jin's Former Residence
Shanghai Zhangming Architectural Design Firm
Shanghai Automotive Industry Corporation
Nanjing Mochen Image Media Co.,Ltd.

上海武康路建筑地图
Shanghai Wukang Road

乔争月 张雪飞 著
Michelle Qiao Zhang Xuefei

同济大学出版社
TONGJI UNIVERSITY PRESS

1923年福开森路的住宅　　A house on rue Ferguson, 1923

"There was nothing unusual about the two pairs of wooden gates we saw today on Route Ferguson, near to the Route Maresca corner: nor was there anything attractive about the bamboo fence. We have passed the place hundreds of times and we never suspected that the somber fence was a line of demarcation between a public road and such a charming garden. It is another small garden upon which great care and attention has been bestowed and we came away with the impression that the present occupier of the premises will continue the good work and make the place still more attractive by adding that personal touch without which even a well designed place can become just ordinary. An enclosed verandah, the inside colour scheme of which blends perfectly with the foliage of the surrounding trees, is covered with creepers; wistarias are arranged at either end and a vine in the center. A low grass-covered bank extends along the front of the verandah; this is planted with Japanese junipers, roses and violets while on a terrace near by we noticed *Asparagus Sprengeri* and euonymus growing in green glazed pots. There are not many flower beds in the garden but the few there are, are attractive: one is full of vinca while another at the base of a bird bath, is planted with torenia, petunia and portulaca. In the bath, we noticed some life-like porcelain doves, looking so natural. A line of cedar trees divides the garden into two sections and beautiful evergreen trees are planted in the boundary shrubberies. We saw numerous cedars, Japanese junipers, pines aucuba, and podocarpus: the various shades of green

1923 年福开森路的住宅　　Houses on rue Ferguson, 1923

are most effectively arranged while an occasional rockery stone breaks the monotony. There is a charming section north of the croquet lawn and close to two fine old weeping willow trees: there, an ornamental pool has been arranged with an island planted with bamboo plants in the center. A bridge connects the island with the mainland and rocks are protruding from small mounds along the mainland edge: some dwarf trees are growing on these mounds. We particularly liked the effect produced by groups of porcelain ducks: one 'mammy' was taking her brood over the bridge on a visit to another family amongst the bamboo while a fluttering duckling, appeared to be annoyed with its 'mammy' for sitting on the mainland and not joining the visiting family. The "black sheep" was there also: a lone duckling was seen far away on the opposite side of the pool, the whole scene was very realistic. We completed our tour of the garden by walking along an enchanting pathway along the boundary, behind the shrubberies and past young walnut trees, azaleas and other shrubs. We have never seen such a quantity of expensive trees growing on such a small area; but their arrangement makes the garden appear much larger than it really are! A garden "god" has a building all to itself and if it was placed there to ensure 'peace and plenty' it has done a good job of work as far as the garden is concerned."

—— **W.J.K**, *North-China Herald*, **September 18, 1940**

今日我们在福开森路靠近赵主教路（今五原路）看到两扇木门和竹篱笆，并无特别吸引人之处。我们路过这里很多次了，从来没发现这黯淡的篱笆居然是马路和一个迷人花园之间的分隔。

这是一个被精心打理的花园。围合的游廊两侧覆盖着爬山虎和紫藤，中央是葡萄树。游廊内侧的色彩与周边树叶的颜色完美交融在一起。游廊的前部由绿叶掩映，种植着日本桧树、玫瑰和紫罗兰，而露台则摆放着绿色釉花盆，种有天冬草和卫矛。

花园里的花坛不多，但都非常好看——一个种满长春花，另一个在鸟池底部，种有蓝猪耳、矮牵牛花和马齿苋。鸟池里有栩栩如生的陶制鸽子，看上去非常自然。一行香柏把花园分成两部分，美丽的常绿树种在分界的灌木丛那里。我们看到数不清的香柏、日本桧树、松树、桃叶珊瑚和罗汉松。深浅不同的绿色被有效合理地搭配好，假山石生动点缀其间。

1924 年的武康大楼　Normandie Apartments, 1924

槌球草坪北面，在两棵美好古老的垂柳边，有一个带小岛的水塘，岛中央种着竹子。通向小岛有一座桥，桥边的土堆上长着矮乔木。

我们特别喜欢一群陶制鸭子营造的效果：仿佛一个鸭妈妈带着一行小鸭子过桥去探访竹林中的朋友。有一只小鸭坐在岸边扇动翅膀，好像被妈妈惹烦了，不高兴参加集体活动。岸对面也有一只这样落单的"败家子"小鸭。整个场景营造得栩栩如生。

最后我们顺着分界线灌木丛后一条动人的道路漫步，走过年轻的核桃树、杜鹃花和其他灌木。我们从来没有看到过如此多数量的名贵树木种植于这么小一块地方，但植物的巧妙安排使花园看上去比实际面积大了许多，有很多美妙的景致。园里还有一座供奉花园神的建筑，也许是为了保佑这一方天地的"宁静与丰饶"，看来真是起作用了。

1940 年 9 月 18 日，《北华捷报》，一封署名"W.J.K."的读者来信

目录

序言 ················· 16
城市山林 ············· 22

武康路变迁 ··········· 26
福开森与福开森路 ····· 28
百年前的"田园城市" ··· 34
武康路的树 ··········· 38
幸运武康路的"微整形手术" ··· 44

上海武康路建筑地图 ··· 48

WK-4
民国海归的摩登别墅 ··· 50

WK-12
谭师的庭院 ··········· 56

● WK-40-1
让历史哗然的恬静小楼 ··· 60

WK-40-4
绿树荫浓的颜氏旧居 ··· 64

WK-67
低调的陈公馆 ········· 70

WK-99
风笛悠扬的大班洋房 ··· 76

WK-113
巴金故居的笑与泪 ····· 80

FX.W-147
柯灵的精巧之家 ······· 86

FX.W-193
英国漫步专家的乡村别墅 ··· 92

WK-115
云朵掠过的金领公寓 ··· 98

HN-262
庭院深深的湖南别墅 ··· 104

WK-129
空留回忆的西班牙小屋 ··· 108

WK-240
大洋行设计的"小熨斗" ··· 112

WK-117-1
中西合璧的银行家别墅 ··· 116

WK-378
武康路上的"城中村" ··· 120

WK-390
酝酿桑塔纳的地中海别墅 ··· 126

WK-393
武康路驿站的双重空间 ··· 130

WK-395
水晶与玫瑰 ··········· 136

HH.M-1843
宋庆龄的船形别墅 ····· 140

HH.M-1850
一枚上海符号 ········· 146

山林城市 ············· 152
推荐阅读 ············· 159
图片来源 ············· 161

WK = 武康路　FX.W = 复兴西路　HN = 湖南路　HH.M = 淮海中路
● WK-40-1：武康路 40 弄 1 号

Contents

Foreword ········· 18

A City Garden ········· 24

The Development of Wukang Road ······· 26

Dr. Ferguson and Route de Ferguson ····· 30

A Garden City in Shanghai ············· 36

Trees on Wukang Road ··················· 40

Microsurgery for a Road ················· 46

Shanghai Wukang Road ·········· 48

  WK-4
  Family Tree in a Modern Villa ·········· 50

  WK-12
  Tan's Courtyard ····················· 56

• WK-40-1
  A Quiet, Renowned Villa ············· 60

  WK-40-4
  Dr. Yan's Home ······················ 64

  WK-67
  A Politician's Home ·················· 70

  WK-99
  Pipe Music from Taipan's House ········ 76

  WK-113
  Tears and Joy in Ba Jin's Residence ····· 80

  FX.W-147
  A Tiny, Smart Home ················· 86

  FX.W-193
  The Country Villa of a Country Walker ··· 92

  WK-115
  Shanghai's Most Unique Apartment House ··· 98

  HN-262
  Villa with a Deep Garden ············· 104

  WK-129
  A Spanish House of Memories ········· 108

  WK-240
  A Mini "Flat-Iron" ··················· 112

  WK-117-1
  A Financier's Villa ··················· 116

  WK-378
  A Chinese "Urban Village" ············ 120

  WK-390
  A White Villa ······················· 126

  WK-393
  A Historic Stop on Wukang Road ······ 130

  WK-395
  Radium and Roses ·················· 136

  HH.M-1843
  Madam Song's Ship-Shaped Villa ······ 140

  HH.M-1850
  A "Shanghai Symbol" ················ 146

A Garden City ······················ 154

Recommended Readings ············ 159

Image Source ····················· 161

WK = Wukang Rd.   FX.W =Fuxing Rd.(W)   HN =Hunan Rd.   HH.M =Huaihai Rd.(M)

• WK-40-1 : No.40-1 Wukang Rd.

序言

111年前辟筑的福开森路，在75年前当上海的道路将所有的洋名按照全国各地的城市重新命名时，这条路选择的是浙江武康。虽然这个街区，这条马路，这些建筑和住在这里的人可以说与武康浑身不搭界，但是经过还谈不上沧海桑田的发展历程，武康路的氛围就为这条马路增添了人文色彩。武康路被命名为"国家历史文化名街"后，成为近年来上海最受关注的街道，其关注程度甚至远远超过当年上海最热门的商业街——南京东路和淮海中路。

如果按照《上海市城市总体规划（2017—2035年）》中关于城市的诗意描述：建筑引导人们去阅读，街区宜于漫步游憩，市民讲究诚信文明，城市散发着暖心的温度，也许可以毫不夸张地说，武康路就是最佳样板。武康路犹如建筑博物馆，这里有许多由著名的中外建筑师设计的各种风格的历史建筑。当人们徜徉其中时，不经意中会邂逅法国式、英国式、西班牙式、意大利式以及历史上不可名状的各式建筑。中国建筑师董大酉、奚福泉、谭垣、范能力、李宗侃等，匈牙利建筑师邬达克、英国建筑师思九生和公和洋行、法国建筑师王迈士、俄国建筑师罗平等都在武康路留下了他们的传世之作。

这里的每一座建筑都有故事，有的壮烈豪放，有的则凄楚委婉，有的如大漠孤烟，有的则如小桥流水，它们汇成武康路的风采。这里的广场、街道、花园、阳台、台阶、墙面、窗户、花坛都是历史的见证，这里的每一个院落就是一篇散文，住在这里的每一个人都是一部历史。武康路并非宏大叙事，并不靠体量，靠伟大的纪念性，或者靠浮华去赢得人们的赞誉。这里是人们日常生活的空间，让人们在这里休憩、嬉戏、约会、读书、聊天、闲坐、晒太阳、发呆。它能启发人们的创意，让人们更热爱生活，热爱我们的城市。

当我们读《上海武康路建筑地图》时，仿佛看到饱经沧桑的阳台，依稀听见绿荫丛中那些窗户后面回荡的琴声，听见客人来访时的门铃声，闻到花园里丁香的芬芳，触摸到墙面上那粗糙的拉毛粉刷或河卵石的肌理，望见冬天夜晚窗户中那一束温暖心灵的灯光。如果说喷泉、松树和广场象征着罗马，让音乐家为之谱曲，而武康路秋天的梧桐落叶、冬天映在墙面上的斑驳阳光、春天透绿的篱笆和夏天树影婆娑的庭院也值得人们去谱写乐曲，本书的作者用专篇来描写作为"城市山林"的武康路的梧桐树，实为知音。

武康路也不只是一条马路，武康路串接了华山路、安福路、五原路、复兴西路、湖南路、泰安路、兴国路和淮海中路，这些马路共同烘托了武康路，因而作者也把叙事引向复兴西路和淮海中路的建筑。这本小册子的作者已经是描述上海的历史建筑和历史街区的专栏作家，已经是研究上海历史建筑的专家，发表了诸多文章和专著，对认知上海的历史建筑有很大的贡献。她撰写的《上海外滩建筑地图》广受好评，至今不衰。我也经常向她求教，请她帮助考证。《上海武康路建筑地图》描述了她作为武康路居民的观察、考证和体会，她用文字和摄影向我们倾诉历史。

期待看到她有更多关于上海的文章和专著问世。

2018 年 7 月 10 日

Foreword

It has been 75 years since the 111- year-old Wukang road was renamed after Wukang, a county in Zhejiang Province from Route de Ferguson when all the foreign-named streets in Shanghai's foreign concessions were to be renamed after Chinese cities in 1943. It seems as if the road, its neighborhood, the buildings along the road as well as the residents have little to do with the city Wukang. Still, a special ambience has emerged thanks to the unique cultural and social elements which have been generated in its short but eventful history. Having been awarded the title of national historical and cultural street, Wukang road has recently become the most noteworthy street in Shanghai. Its popularity has even surpassed Shanghai's two most prominent shopping streets—Nanjing Road E. and Huaihai Road M.

The poetic description included in *Shanghai's 2017-2035 General Urban Plan* describes a city where buildings are readable, streets are nice for walking and relaxing, citizens are honest and civilized while the city is cozy and warm. Wukang Road could be its best example of this character. The road is like an architectural museum that exhibits historical buildings of all styles designed by renowned foreign and Chinese architects. Strolling along the road, one can easily glimpse buildings in styles such as French, British, Spanish and Italian and even some historical styles that seem to defy definition. An impressive constellation of architects, including Chinese architects Dayu Doon, Fohjien Godfrey Ede, Harry Tam, Chi Chen Architects, Michael Li Tson-Cain, Hungarian architect Laszlo Hudec, British architects Stewardson & Spence, Palmer & Turner, French architect Max Ouang, Russian architect Gabriel Rabinovich all have works included on the street.

Here every building has its own story to tell. Some stories are grand and heroic while others are subtle and tragic. Some are as poetic as lonely smokes swirling in the desert while some are remind the stroller of small bridges over flowing streams. All of these styles converge to generate the rare charisma of Wukang Road. All the squares, the streets, the gardens, the balconies, the staircases, the walls, the windows and the flower terraces here have born witness to this area's unique history. Every yard is an essay and every resident here is a piece of that continuing history. Wukang Road is not about those grand narrations and never wins applauds in terms of its impressive scale, grand commemoration or superficial vanity. It is rather a space of everyday life, where people relax, play, date, read, chat, sit idly, sunbathe or simply day dream. This is a road to inspire creativity, to arouse people's love for life and our city.

When we read *Shanghai Wukang Road*, it is as if we can see those withered balconies, enjoy the random piano music resounding behind those windows concealed behind green shades, hear the door bells ringing for a visiting guest, smell the clove fragrance in the garden, touch the textures of those rusty stucco or smooth oval walls, and catch a glimpse of soothing the light from a window on a winter's night. If fountains, pines and squares are the symbols of Rome which attracted musicians to compose music, Wukang Road's fallen plane leaves in autumn, the mottled lights and shadows on the walls in winter, green sceneries penetrating through bamboo fences in spring and those lovely yards with shadows of swinging trees in a summer breeze are also worth a try for musicians to compose music for. The author really understands this as she specially dedicated an article about plane trees on the Wukang Road neighborhood which is regarded as a city garden.

Wukang Road is not a single road. It has linked Huashan Road, Anfu Road, Fuxing Road W., Hunan Road, Tai'an Road, Xingguo Road and Huaihai Road M., all of which have further set off the beauty of Wukang Road and explains why the author has chosen to also write about architecture on Fuxing Road W. and Huaihai Road M. The author is more than a columnist on Shanghai's urban history and historical buildings but is also an expert in studying Shanghai's historical buildings. Her efforts include articles and books which have contributed greatly to promoting the city's historical architecture. Her book *Shanghai Bund Architecture* is highly-regarded and enjoys a lasting popularity. I often turn to her for reference help. Telling the history through words and photos, *Shanghai Wukang Road* is a collected record of her observations, research and feelings as a resident of Wukang Road.

I expect her to write more articles and books about Shanghai.

Zhang Shiling

July 10, 2018

城市山林

2009年的落叶季，我搬到武康路的一条弄堂。当时静悄悄的武康路还没有成为"网红"，朋友问起新家，要说住在"上海图书馆附近"。

那个落叶季后，武康路的人气渐旺，游人如织，她在朋友圈的出镜率也越来越高。这变化让我好奇，是什么让这条安静低调的小马路变成"网红"？我想弄明白。

武康路的底子好。长1 183米的武康路辟于1907年，原名福开森路，因美国传教士福开森命名。1914年法租界第三次扩张，武康路所在区域被划入法新租界。公董局以打造宜居社区为目标进行规划，从路边建筑的营业类型到行道树间距都有管理章程明文规定。规划可能受到当时的国际先进规划理念——"田园城市"的影响。1943年，福开森路在汪伪政府时期被改名为"武康路"。据说，路名由汪精卫亲自修改，他认为福开森路虽在大都市，但其环境和氛围与他曾到访的浙江武康（今德清县）莫干山十分相似。

中华人民共和国成立后，武康路一带由于规划合理、房屋品质好，依然是上海政界、工商界、文化界和艺术界人士汇集的地方。这条路的人脉和文脉得以延续，唯独路况，逐渐跟不上城市发展的速度，日显窘迫。

时光流转到2007年，武康路幸运地被选为全市第一条风貌街道保护试点，由徐汇区负责进行保护性规划和综合整治。历史仿佛轮回，试点工程也吸取国际先进经验，邀请同济大学沙永杰教授作为总设计师，牵头协调、统一风格。设计团队尽量不着斧斤，用与武康路气质契合的天然材质修缮围墙和大门等风貌要素。"微整形手术"历时三年，武康路就这样默默地变美了。

2010年上海世博会期间，武康路被选为展示上海城市保护的参观点；

2011年又入选"中国历史文化名街"。2013年，武康路成为上海最早试点"落叶不扫"的道路之一。洒满金色落叶的历史街道成为一道风景，武康路的知名度和吸引力大大增加。

与此同时，徐汇区政府在武康路393号开设了"老房子艺术中心"。民间社会力量也乘势而为。由民间资本开发、采用都市"城中村"设计理念的武康庭里，开出一家家富有格调的餐厅和咖啡馆，这条风貌道路活色生香起来。

1936年版的《大上海指南》提到武康路所在的区域，"布置之整洁，空气之新鲜，为全市之冠。因此，外侨及中等以上华人多寄居于此，风景有如乡村。闲来无事，街头散步其乐也融融"。

城市山林，都市田园，中外交汇，名流云集，这就是武康路的气质。

"网红"武康路，源于百年前法租界公董局精心打造了一条人性化的街道，源于十年前上海市政府及徐汇区政府一次较为成功的风貌道路保护尝试，也许更源于都市人内心对美好生活的向往。

无论百年前福开森路第一批洋居民，1936年徜徉街头散步的民国人，还是今天在上海生活的都市人，到位于城市中心但风景有如莫干山林的武康路漫步，都是一件其乐融融的事。

乔争月
2017年11月 于武康路月亮书房

A City Garden

When I moved my home into a hidden lane on Wukang Road in 2009, the rejuvenation project for this historical road by the local government had just been completed. While Wukang Road had become much cleaner, neater and more beautiful, it remained quiet and unknown. Before the renovations, I often had to clarify my address to people by adding "near the Shanghai Library."

Now Wukang Road has become one of the city's star streets. Couples can be seen posing for their wedding photos, models for fashion shoots, clusters of tourists strolling around for nice coffees, ice creams and old buildings. I don't need to explain where I live any more, but, on occasion, must explain how this quiet little road becomes so popular.

Stretching 1183 meters, Wukang Road was built in the 1890s. First a nameless road, then it was expanded and officially named Route de Ferguson in 1907 to honor the contributions of the American missionary John Calvin Ferguson.

The road was included in the former "new French Concession," a vast area that the former French Municipal Council gained by expanding its concession westward to Huashan Road in 1914. The area was planned as a high-end residential zone to accommodate the city's growing wealthy population. Later, its urban planning mirrored the then popular British concept of "Garden City" in many ways.

Still crisscrossed with farm fields, villages, graveyards and small rivers in the early 1900s, the neighborhoods surrounding Rue de Ferguson quickly grew to be an idyllic, convenient community graced by garden villas and apartment buildings during the 1920s and 1930s.

In 1943, Route de Ferguson was renamed Wukang Road presumably by Wang Jingwei, the head of the Japanese Puppet Government. It is said Wang discovered a resemblance between the environment and ambience of Route de Ferguson and Mount Moganshan in Wukang County, Zhejiang Province.

Some experts even regard the Wukang Road neighborhood as "the only well-planned, high-quality residential area in old Shanghai," where wealthy expatriates and Chinese merchants, politicians and celebrities had chosen to live.

However, before the rejuvenation project kicked off in 2007, the road seemed to wane away and lag behind Shanghai's speedy urban development. Concerns mounted among some and, that year, Wukang Road was very lucky to be selected from the city's 64 first-class historical streets for the first pilot revamp project.

The rejuvenation project was regarded as the city's first attempt to restore and revive an historical street. History seemed to repeat itself to some extent as the project also drew inspiration from advanced international experiences and commissioned Tongji University professor Sha Yongjie, a Harvard University graduate, as the chief planner.

He coordinated different institutions to unify styles and his design team used natural materials for restorations that seemed compatible with the idyllic style of this neighborhood. Wukang Road regained its natural beauty through this three-year "micro-surgery". It proved to be so successful that it won national awards and the government planned to copy "the Wukang Road model" and apply it to other historical streets.

In 2011, Wukang Road was named a national historical and cultural street by the State Administration of Cultural Heritage. In 2013, Xuhui District tried a "no-sweep of fallen leaves" experiment on the tree-lined Wukang Road which had a great effect in attracting people to enjoy the golden fallen foliage in autumn.

As the chic restaurants and cafes popped up on both sides of the street, the clever district government set up a visitor center on the road to attract more visitors.

The 1936 book *Great Shanghai Guide* says the Wukang Road neighborhood boasted "the city's neatest environment and freshest air."

"Therefore expatriates and middle or upper class Chinese chose to live here. The scenery is like countryside. During leisure times it's a pleasant walk to stroll around the streets."

Although, the instant popularity of Wukang Road could definitely take credits from the efforts of the former French municipal council to create a pedestrian-friendly street some 100 years ago and the efforts of the Shanghai government to revamp a historical street 10 years ago, ultimately it was the result of modern people's craving for a better city and a better life.

Today it's still a pleasure to stroll down Wukang Road, which sits in downtown Shanghai but carries the natural beauty of Mount Moganshan.

Michelle Qiao
November 2017
Moon Atelier, Wukang Road

武康路变迁：20 世纪 10 年代　　The Development of Wukang Road, 1910s

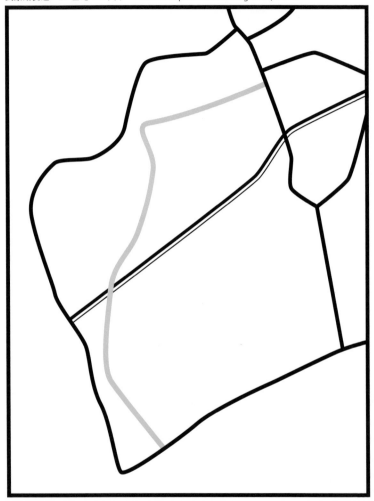

▬ 福开森路　　　▬ 宝昌路（今淮海中路）
Route de Ferguson　　Avenue Paul Bruna (Today's Huaihai Road M.)

武康路变迁：20 世纪 20 年代　　The Development of Wukang Road, 1920s

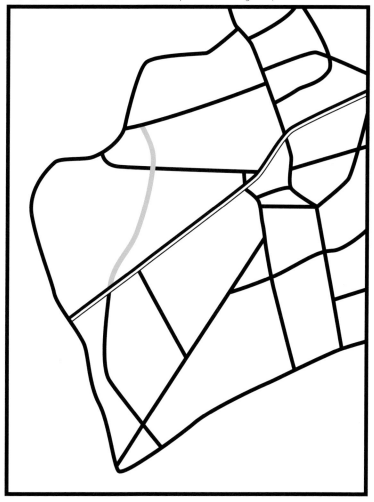

▬　福开森路（今武康路）
　　Route de Ferguson (Today's Wukang Road)

▬　霞飞路（今淮海中路）
　　Avenue Joffre (Today's Huaihai Road M.)

福开森与福开森路

武康路原名福开森路,因为"该路系美国福开森所建筑"。

1924 年出版的《上海轶事大观》记载,美国传教士福开森(1866—1945 年)任南洋公学监院后,因以公学附近交通不便,捐银筑马路一条。"造成后,初无确实名称,后经该处居民即以先生之名为路名,谓之曰福开森路,至今未之改云。"

同济大学钱宗灏教授认为,这条路如此命名,更是因为福开森是中国近代历史上的重要人物,曾经在中国与西方国家的政治文化交流中发挥突出作用。

他提到,"福开森旅华前后计 57 年,在中国度过了自己一生中最重要的时光。他集传教士、教育家、报业大亨、政治顾问、慈善家和文物专家等头衔于一身,对中国近代政学两界颇有影响,对传播中国文化有相当的贡献。"

这 57 年的经历异常丰富。福开森担任过晚清和民国政府的顾问,创办《新闻报》,参与创办两所大学——南京汇文书院(金陵中学和南京大学前身)和上海南洋公学(上海交通大学前身)。他是中国文物收藏家,曾是故宫文物鉴定委员会唯一的洋委员,著有多本有关中国文物和古代艺术的专著。他的藏品陈列于纽约大都会博物馆和南京大学考古与艺术博物馆。

1937 年 10 月 27 日,英文《北华捷报》在头版刊登文章,纪念福开森先生来华 50 周年。

> 1887 年 10 月 25 日,福开森夫妇从旧金山抵达上海。一年后,他赴南京旅居 10 年,此后又返回上海直到 1911 年去北京。福开森先生在华这 50 年间,历经中国最重要的三个城市。

福开森先生对中国事务热切关注。无论新闻工作，还是对中国艺术珍宝的权威研究，他都得到高度评价。他对中国教育所做的工作也得到政府的肯定，所以回顾在华旅居的漫长岁月，他深感自己的工作和观点都能得到欣赏和认同。

福开森先生的良好体魄说明中国气候有益身体健康，能历经多年沧桑依然保持体力充沛。庆祝福开森来华50年的庆典也特别提及已故的、曾经深受爱戴的福开森夫人。10年前，在纪念福开森先生来华40年之际，我报曾评论道：

"福开森先生与几乎中国所有军政界的大人物关系亲密，而各种打击和幻灭从未让他失去对中国人民的信心，这非常少见，难能可贵。他那非凡的风度、好脾气和判断力，让他在中国人和在华外国侨民里享有极高的威望。"

而这段致敬词今日（1937年）也不过时。宽广的胸怀和对人性的信念造就了这位真正的政治哲学家。福开森博士值得尊敬。朋友们希望他继续工作10年，享受旅居中国更多的乐趣。

Dr. Ferguson and Route de Ferguson

While most streets in the former French Concession were named after prominent Frenchmen, Route de Ferguson (Wukang Road) stands out as an exception.

The name originated with the American missionary John Calvin Ferguson, who was said to have built the road to make traveling to Nanyang College (now Shanghai Jiao Tong University) more convenient for the faculty who lived in homes downtown, according to the 1924 book *A Comprehensive View of Shanghai Anecdotes*.

Tongji University professor Qian Zonghao said naming a road after Ferguson showed he was an important figure in modern Chinese history, who had done a terrific job promoting cultural and political links between China and Western countries.

"Ferguson had spent the most important 57 years of his life in China. He was a missionary, educator, newspaper tycoon, governmental advisor, philanthropist and antique expert who had a positive influence on the political and academic circles in modern China while contributing to spreading Chinese culture abroad," Qian says.

Ferguson lived a full life during the 57 years he spent in China. Born in Canada in 1866, Ferguson later moved to the United States with his family and graduated from Boston University in 1885. One year later he was sent to China with his new wife as a missionary.

In China he helped found two modern Chinese universities—Nanking University in Nanjing and Nanyang College in Shanghai. He purchased and successfully managed the Chinese newspaper, *Sin Wan Bao*, for 30 years and acted as an advisor for the Chinese government.

Ferguson had also been one of the earliest Western scholars to collect Chinese art. He was the only foreign council member of a committee to catalog the imperial palace's art collection after the fall of the Qing Dynasty in 1911.

His two pioneering indexes of writings on Chinese art in the 1930s—one for paintings and the other for bronzes — were basic references for the next generation of scholars. Taking advantage of his knowledge and connections, he had also helped the Metropolitan Museum of Art in New York acquire works by Chinese artists.

In 1934, Ferguson donated much of his own collection, including bronzes, scrolls, paintings and jade items to Nanking University. He stayed in Beijing until the end of 1943 and died in the US in 1945 at the age of 79.

On October 27, 1937, the *North-China Herald* published an article to celebrate the 50th anniversary of his arrival in China.

> "Fifty years ago on Oct. 25, 1887, Dr. and Mrs. John C. Ferguson arrived in Shanghai from San Francisco. After a year in Shanghai as a language student, John Ferguson went to Nanking where he remained ten years; then he returned to Shanghai and until 1911 he made his home in this city. From 1911 he and his wife moved to Peking where they have lived ever since. Thus Dr. Ferguson's fifty years in China have been spent in three of China's chief cities. As an eager student of Chinese affairs, as a journalist and as an authority on Chinese art treasures Dr. Ferguson is highly esteemed. His services to Chinese education have been specially recognized by the Government and he can look back on a long period of residence in a country where he has made many friends with the comforting reflection that his work and views are appreciated. His physical fitness testifies to the salubrity to China's climate and to his own stamina in facing the vicissitudes of years which have been of such momentous importance to China's national development....Ten years ago, in commenting on Dr. Ferguson's survey of forty years' experience of China this journal remarked:

民国元年南洋公学同学会合影
Students of Nanyang College in the first year of Republic of China (1911)

南洋公学中院
The middle school building of Nanyang College

Dr. Ferguson is probably one of the few men who have been intimately connected with all of China's biggest political and military leaders who has not allowed the inevitable discouragements and disillusionments to rob him of his confidence in the Chinese people. His remarkable poise and good temper, sound judgement and tact, have made it possible for him to exert an influence which has been exceptionally wholesome for both China and the foreigner resident in China.

That tribute has not become musty with the passage of the years. It endowed its subject with that breadth of vision and faith in human nature which must necessarily form part of the equipment of the true philosopher in politics. So Dr. Ferguson has remained to be saluted as he records another decade in a career which his friends hope will have more to give him in enjoyment of happiness in China. "

百年前的"田园城市"

20世纪初,福开森路吸引了许多中外精英居住生活,这与法租界为建设宜居社区所作的努力有关。

上海法租界于1849年建立,后来历经三次扩界,1914年,福开森路所在的街区在第三次扩界时被并入法租界。这最后一次扩界的区域常被称为"法新租界"。

研究上海法租界城市管理的上海社会科学院学者陆烨发现,法租界公董局很早就想在中西部新扩地区建设一个高级的住宅区。20世纪初,法租界东部已聚集繁荣的商业和行政部门,所以中西部地区没有太多政治和经济方面的需求。而公董局的收入依赖于房地产的地税和房捐,中外地产商们因此鼓动公董局在租界的中西部区域建设高级住宅区,以舒适的居住环境吸引投资者,创造商机。

为打造宜居社区,公董局对法新租界进行科学细致的城市规划和管理,从营业类型限定、行道树绿化、摊贩管理到垃圾清运都制定了周密的管理章程。住宅的式样被限定为欧式,因为当时认为中式建筑是不安全和不卫生的。

1938年的《整顿及美化法租界计划》将法租界分成几个区域,福开森路被划入有着高标准定位与严格管理的"A区"。这一区域内的住宅建筑必须具有相当的墙身间宽度,附有暖气和卫生设备,是当时高级住宅的标配。

法新租界的规划可能还受到当时国际最新的城市规划理念——"田园城市"的影响。

1898年,原英国议会下院速记员霍华德出版名著《明日的田园城市》。他在书中写道:大城市收入高,但物价高、空气受污染;乡村空气清新,可是工资低,又缺乏娱乐和公共设施。城市和乡村各有优缺点,应该想办法集合城乡优势,打造既生活便利、公共资源丰富又充满自然风情的"田园城市"。1903年,他开设"田园城市有限公司",先后在伦敦附近建设了两座田园城市——莱奇沃思和韦林。后来霍华德的"田园城市"并未得到广泛的复制和推广,但其理念却在国际城市规划界产生了深远影响。

20世纪初的上海开始出现拥挤和污染等大城市病,居住环境恶化,英文报纸刊登了关于"田园城市"的介绍和讨论。陆烨发现法租界的规划与田园城市理念相同之处很多,如绿化用地和污染控制。也许,福开森路一带的规划建设或多或少地受到这种理念的影响,形成以竹篱围墙、绿植和天然材质立面为特征的城市空间。

今天在武康路漫步,是否感觉身在一座田园城市呢?

A Garden City in Shanghai

Wukang Road was and is still one of the most popular areas of the city due to its well-planned neighborhoods.

The road is in the former "new French Concession," a vast area that the French Municipal Council gained by expanding its concession westward to today's Huashan Road in 1914. It was the last time the concession expanded. The area was planned as a high-end residential zone to accommodate the city's growing population of wealthy individuals.

According to PhD researcher Lu Ye from the Shanghai Academy of Social Sciences, the council had a strong vision for establishing an upscale residential area in the middle and western French portions of the concession after the 1920s. As the eastern part of the concession had already attracted commercial and governmental institutions, the council valued the residential environment over commercial interests when planning the new area. Chinese and foreign developers also promoted this idea for business opportunities brought by a comfortable living environment.

The French Municipal Council then executed strict management to create a clean, tidy neighborhood. They collected a hygiene tax and then established a committee to manage area businesses.

More importantly, the committee classified businesses into three major categories according to the degree of damage they would cause to the residential environment. The dirty chemical industry was in the most dangerous category, while workshops and food stores were next, followed by the "nearly harmless" cafes, bistros and clinics.

Under strict management, businesses that would bring noise and foul odors, or affect hygiene and safety were removed from residential areas. Meanwhile, service businesses and cultural institutions were encouraged in the residential areas to make life more convenient and pleasant. Soon, the residential areas became sprinkled with hospitals, schools, clubs, libraries, theaters and churches.

After these ideas were implemented, the former French Concession was divided into a financial/trade zone in the east, upscale residential zones in the middle and west including Wukang Road, and a commercial zone along Avenue Joffre (today's Huaihai Road).

The urban planning in the former French Concession may have drawn inspiration from some forward thinking of the time, including UK'S "Garden City" concept.

Ebenezer Howard's "Garden City" concept, published at the turn of the 19th and 20th centuries, focused on reconciling the increasing tension between industrialization and comfortable living. He defined a garden city as a town designed for both healthy living and industry with a surrounding rural belt.

Shanghai's urban environment at the time was being increasingly harmed by industrialization along with the growing population early last century. Local English newspapers seemed to have introduced the "Garden City" concept as a response to these mounting concerns. It's possible that urban planning in the Route de Ferguson neighborhood had been influenced at least indirectly by this concept, resulting in a more tranquil urban space featuring bamboo fences, tree-lined streets, lush garden villas and an abundant use of natural materials on buildings.

武康路的树

武康路最动人的美,就是一幢幢老房子在林荫枝影间若隐若现。无论春夏荫浓,秋冬落叶,晴日树影婆娑,还是挂着雨珠的油绿树叶,高大茂盛的行道树为这条街道增色不少。

上海市绿化和市容管理局资料显示,武康路的行道树种植晚于道路建设,具体年代无法考证,现有行道树250株,树木高度平均约13米,人行道和车行道的绿化覆盖率分别是92%和82%。

武康路的树种既有乡土植物枫杨,又有西人后期种植的悬铃木,中西合璧。

传统中国城市并无栽种行道树的习惯。据《上海园林志》记载,上海行道树的历史可以追溯到1865年,英美租界当局在外滩江边种植树苗。几年后,法租界也开始在外滩码头种植,后来扩展到今天的徐家汇路、建国西路和天平路一带,树种以悬铃木为主。

法租界公董局档案记载,1887年从法国购回250株悬铃木苗和50株桉树苗,次年2月试种,结果悬铃木的生长远比桉树好。此后公董局多次从法国成批购买悬铃木树苗,到1939年法租界有行道树1.29万棵。

1925年3月,《北华捷报》报道了圣约翰大学教授W. M. 波特菲尔德题为"上海树木"的讲座。教授特别提到引种到上海的"伦敦梧桐"长势良好,是理想的行道树。

"伦敦梧桐"就是悬铃木,其树形雄伟,枝叶茂密,是在世界范围内被广泛种植的行道树之一。悬铃木主要分为一球、二球和三球三种,分别俗称美国梧桐、英国梧桐(即伦敦梧桐)和法国梧桐。上海地区的悬铃木均为二球悬铃木。悬铃木最初由法国人引种在法租界内作行道树,人们看其叶子像梧桐树,又是法国人在推广,就叫它"法国梧桐"。

公董局不仅开办多处苗圃培养种苗,还制定关于路旁植树的管理章程，做出细致的规定，如树间距需保持在 7～10 米。

老上海的外国侨民对植物和环境非常关注。1892 年，博诺·普布利科给《北华捷报》写了一封读者来信，呼吁为了阴凉和健康，上海应多种植和养护好行道树。他提到南京路没有树荫，马路在炙热的夏日午后尘土飞扬。

1925 年 4 月，苏柯仁在《北华捷报》发表了一篇题为《树》文章，认为"一个花园、一座城市或一个国家没有树，就像房子没有家具一样……如果没有树，那人们应该种植，如果有树，那就要好好照料……"。文章还配有一张珍贵照片，展示了瑞金二路一棵高达 40 多米的大树。

一位署名"W. J. K."的读者在 1940 年 11 月 13 日的《北华捷报》上指出，上海的悬铃木是"伦敦梧桐"。他惊讶于悬铃木耐修剪，"它们可以被修剪到只留下树干，但却不会失去活力，次年便会长出三四米长的新枝。"这篇文章也提及悬铃木的毛絮会引发口鼻黏膜的炎症，所幸每年飘絮时间不长，还能忍受。

2013 年秋末冬初，徐汇区尝试将武康路和余庆路营造成"落叶景观道"，观赏体验时间虽不长，给人的感受却非常震撼。2014 年，武康路被命名为"上海林荫道"。而悬铃木自 19 世纪引入上海法租界后，陆续引种推广，作为行道树广植于全国各地。"法国梧桐"的说法也传遍了全国。

上海行道树的种植虽源于租界，但后来华界的行道树也逐渐增多，并有早期研究出版。

1928 年，张福仁在《行道树》一书中写道："故吾人厕身欧美都会，见其道路广阔，市肆整洁，树木繁茂，绿荫缤纷，不觉心旷神怡，尘嚣顿忘，不辨其为山林城市与城市山林也。"

今日武康路，也有城市山林的气质。

Trees on Wukang Road

Today, one of the most attractive scenes on Wukang Road are the majestic trees that run along much of its length, half hiding the many historical buildings on either side. In the summer, these trees create plenty of shade along the road but it is the magnificent and colorful beauty of its fallen autumn leaves for which the street has become most renowned.

According to the Shanghai Department of Afforestation and City Appearance, the 250 trees along Wukang Road have an average height of 13 meters and were planted soon after the street was constructed in 1907.

The trees provide shade for up to 92 percent of the sidewalk and 82 percent of the roadway. There are two main types of trees — the local species Pterocaryastenoptera and the London plane, implanted from Europe.

Traditional Chinese cities had no habit of planting trees along the streets at the time. The book *Record of Shanghai Gardens* dated the city's history of street trees to 1865 when the municipal council of international settlement first planted trees along the bund. The French grew plane trees on the French bund soon after, which expanded to today's Xujiahui, Jianguo W. and Tianping Roads.

Archives showed that the French municipal council purchased saplings of plane trees and eucalyptus. The latter grew much better in Shanghai so the council continued to import plane trees. The former French Concession had 12,900 street trees by 1939.

On March 1925, the *North-China Herald* introduced a lecture on "Shanghai's Trees" by Professor W.M. Porterfield of St. John's University, who said "The London plane has been introduced and grows well. It is a very fine shade tree."

A hybrid of the Oriental Plane (Platanusorientalis) with the American sycamore (Platanusoccidentalis), the London plane has been widely called the "French parasol" since it was introduced by the French and resembled a Chinese parasol.

The French municipal council not only founded nursery gardens to grow saplings but also began actively managing the city's trees through a serious of regulations. For instance, they regulated the distance between trees from seven to 10 meters, adding they should be planted 1.5 meters from the street and from other objects such as telegraph poles or fountains.

The expatriate community residing in Old Shanghai seemed to have paid much attention to the natural elements in the city, especially its plants and trees.

In 1896 a reader named Bono Publico wrote to the editor of the *North-China Herald* complaining about the "terribly treeless condition" along Maloo (today's Nanjing Road E.) He mentioned "the clouds of dust and the faring setting sun in his eyes" on hot summer afternoons and called on councilors to plant a few trees along the broad part of the Maloo.

On April 11, 1925 Arthur De C Sowerby's article said "a garden, or city, or country without trees is like a house without furnishings, and it beloves mankind as much as possible by these most blessed of nature's creatures. If he has not got trees around him, he should plant them. If he has, he should see to it that they are cared for." The article is published with a photo of a 150-feet old tree outside a little temple in the former French Concession.

Still on this newspaper, W.J.K.'s article on November 13, 1940 mentions that "our plane tree (Platanusacerifolia) is the London Plane and it is said to be a hybrid between the oriental plane and the North American Button-wood (Platanus occidentals)."

He was surprised by "how well the plane tree stands pollarding and hard pruning" which may be pollarded to mere stumps and instead of losing their vitality will grow new branches the following year as high as ten feet in length.

He also mentioned the negative effects of the trees including the "star-like" hairs on young leaves which, when detached in early spring" had tendency to induce a inflame the mucous membranes of the eyes and nose but noted that city-dwellers could "stand a little inconvenience for the sake of enjoying the pleasant shade afforded by the large leaves during the summer".

In the autumn of 2013 Xuhui District experimented with a so-called "no-sweep rule" on tree-lined Wukang and Yuqing Roads so that people could enjoy not only the golden canopies above their heads but also the crispy fallen foliage beneath their feet. The scenery was overwhelming and Wukang Road was named a "Shanghai avenue" in 2014.

Starting from the former French Concession in Shanghai, London plane trees were introduced extensively to other parts of China. The name "French parasol" was widely used around the country.

Although street trees originated in Shanghai's settlement, other Chinese towns and districts were soon to follow as other places planted roadside trees and studied their effect on the surrounding city-scape.

In his 1928 book named *Street Trees*, Zhang Furen wrote that "European and American metropolitan cities feature wide streets, clean shops, flourishing, beautiful trees and lawns which made me relaxed and joyful, free from worries. I cannot help wonder if it's a garden city or a city garden".

Today Wukang Road mirrors a city garden portrayed by the Chinese author.

幸运武康路的"微整形手术"

2005年,上海历史保护工作有一个突破,开始点、线、面铺开。"点"是历史建筑,"线"是风貌道路,"面"是历史风貌区。"点"比较分散,而且不是每座建筑都可以参观;"面"尺度大,人们很难形成整体观感;对大众而言,最容易体味与感知的是"线",也就是144条风貌道路。其中的64条一类风貌保护道路品质最好,永不拓宽,武康路就是其中之一。

2007年,上海市规划管理部门要进一步对这64条道路进行保护性规划,武康路幸运地被选为试点。2007年到2009年间,徐汇区对武康路进行保护性规划和综合整治。2010年上海世博会期间,武康路被选为展示上海城市保护的重要参观点,2011年又入选"中国历史文化名街"。

这次试点工程有一个制度创新,由同济大学沙永杰教授担任总规划师,负责协调和指导武康路各个局部的小型设计,这是中国首次在城市保护工作中尝试由一位总规划师负责。

他和设计团队用与武康路气质吻合的天然材料和装饰艺术风格来保护修缮重要的风貌元素。比如武康路的围墙经过历次修缮看上去敦厚实在,这次根据历史照片选择了竹篱笆和卵石等天然材料,有了起伏变化,好看很多。

沙教授透露，改造工作看上去变化不大，其实工作量非常大，一些细节改动要跟多个政府部门协调。比如垃圾箱、电线、电表箱、广告等，虽然小，但对街道风貌影响很大。又如有的历史建筑正立面甚至安放了避孕套自动售卖机。

这次试点工程还注意听取居民意见，把保护工作和修缮居民老房子的事情结合在一起，例如为武康路400弄的公共空间调整了绿化，加了座椅和景观灯，效果很好。居民们也因此非常支持，帮忙劝说邻居清理违章堆放的杂物，参与营造宜居环境。

沙教授认为，这条路选得很好。"武康路之前是不显山露水的一条路。我们做了一点点应该做的'小动作'，做得比较到位。武康路项目算是天时、地利、人和。"

曾担任上海市规划和国土资源管理局风貌管理处处长的王林教授认为，风貌保护最忌讳的就是"一年一变样，三年大变样"。在她看来："武康路改造过了，但感觉好像没有做过，却又发现一些惊喜。街道的品质提升了，变得优雅精致，但却没有翻天覆地的变化。"

经历这次科学细致的"微整形手术"后，幸运的武康路就这样默默地变美了。

Microsurgery for a Road

The rejuvention of Wukang Road won a state-level award for excellence in urban planning and was named a "national historical and cultural street."

The project's chief designer, Tongji University professor Sha Yongjie, said the idea of not making wholesale changes to a neighborhood was a new urban planning concept in China.

The project, begun in 2007, was completed in 2010. It was regarded as the city's first attempt to revitalize a historical street.

The project resulted from the city's groundbreaking 2005 plan to establish 12 historical areas and 144 historical streets in downtown Shanghai.

The 144 streets include 64 that will never be widened in order to preserve their original appearance and the old buildings flanking the roads. As one of the 64 streets selected, Wukang Road was the first to be revamped because it was a straight-forward residential street.

Before the revamp, the historical road was filled with many eyesores including electric wires, dustbins, milk boxes on walls, roadside shops and signage, which detracted from the street's appearance. Apparently the project wasn't just a matter of traditional urban planning but also a matter of refined city management.

Sha had to coordinate with more than 10 municipal, district or community-level government bodies to improve numerous small details. This included adding roadside trees, burying wires, replacing dustbins, removing ugly shop signs and condom vending machines that were in front of historical buildings.

He was especially impressed with some community-level officials, who helped convince residents of the project's benefits.

The cooperation of residents helped the project go smoothly. In earlier renovation projects, residents were often compelled to move out from old homes. This time, the residents' requirements and advice were respected. And with the architects' professional guidance and a modest increase in funds available, the result was so much better than similar government projects undertaken elsewhere.

When renovating Lane No. 400, Sha's team readjusted green plants, added chairs and lights in public space and repaired the drainage system. The residents were so happy that they then helped persuade a neighbor to clean up an assortment of private belongings occupying public space.

For a variety of designs, the chief designer organized a team of several young architectural faculty members from Tongji University. Each scholar/architect was assigned to design a gate or two, or a section of surrounding walls. Sha supervised the designs to ensure they suited the character of Wukang Road.

In addition, Sha's team also replaced the metal facades of some 1980s buildings with stonewalls or bamboo fences to match the old villas.

Upon completion of the project, the popularity of Wukang Road soared. Trendy restaurants and cafes popped up on both sides of the street. The project was praised as "achieving a great effect without costing a fortune compared with other renovation projects."

It proved so successful that the Xuhui District government has made plans to apply "the Wukang Road model" to other historical streets. There are also plans to convert historical buildings into museums near Wukang Road.

Shanghai Jiao Tong University Professor Wang Lin, former vice director of the Architectural Preservation Committee of the Shanghai Architectural Society, said the project was well executed.

"Wukang Road looks more beautiful than ever, but people think nothing has been done here," Wang said. "Then it seems as though something has been done to improve the atmosphere of the road, but without 'upside down changes,' which we historical preservationists fear most."

Sha said they made some small, proper changes and luckily few mistakes. And Wukang Road is lucky to be the first historical street in Shanghai to experiment with the "microsurgery" that brings out its natural beauty.

WK-4
武康路 4 号
NO.4 Wukang Rd.

WK-12
武康路 12 号
No.12 Wukang Rd.

WK-40-1
武康路 40 弄 1 号
No.40-1 Wukang Rd.

WK-40-4
武康路 40 弄 4 号
No.40-4 Wukang Rd.

WK-67
武康路 67 号
No.67 Wukang Rd.

WK-99
武康路 99 号
No.99 Wukang Rd.

WK-113
武康路 113 号
No.113 Wukang Rd.

FX.W-147
复兴西路 117 号
No.117 Fuxing Rd.(W)

FX.W-193
复兴西路 193 号
No.117 Fuxing Rd.(W)

WK-115
武康路 115 号
NO.115 Wukang Rd.

HN-262
湖南路 262 号
No.262 Hunan Rd.

WK-129
武康路 129 号
No.129 Wukang Rd.

WK-240
武康路 240 号
No.240 Wukang Rd.

WK-117-1
武康路 117 弄 1 号
No.117-1 Wukang Rd.

WK-378
武康路 378 号
No.378 Wukang Rd.

WK-390
武康路 390 号
No.390 Wukang Rd.

WK-393
武康路 393 号
No.393 Wukang Rd.

WK-395
武康路 395 号
No.395 Wukang Rd.

HH.M-1843
淮海中路 1843 号
NO.1843 Huaihai Rd.(M)

HH.M-1850
淮海中路 1850 号
NO.1850 Huaihai Rd.(M)

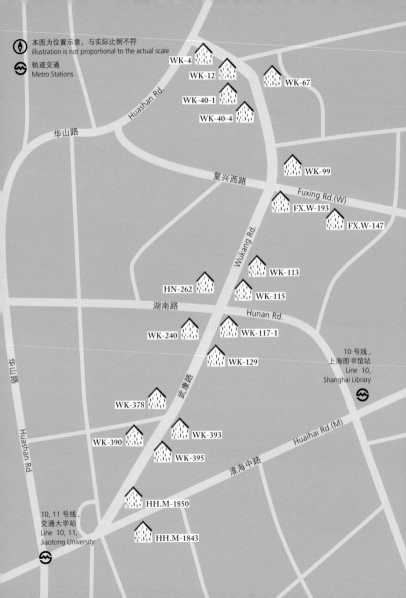

WK-4
民国海归的摩登别墅
The Family Tree in a Modern Villa

　　武康路4号是一座米色的摩登别墅,外观简洁流畅,大大的玻璃窗,没有一丝多余的装饰。小楼与两位民国"海归"有关。

　　一位是昔日主人蔡声白。1919年,他留美回国,乘船抵达上海时,未婚妻莫怀珠家已派汽车在码头迎候。次年,他与这位丝业大王莫觞清的千金结婚。莫觞清对西方文明很有好感,选择这门亲事也是看重蔡氏留美学生的背景。后来他让这位能干的女婿担任旗下美亚丝绸厂的经理。

　　这个决定非常明智。蔡声白就读的理海大学在美国宾州伯利恒市,那里正是"科学管理之父"泰勒曾大力推动科学管理实验的地方。蔡声白活学活用,用"泰勒法"管理传统丝厂,通过制订合理的日工作量、统一技术标准和实行计件超产奖励来提高生产效益,取得成功。

　　1930年,蔡声白邀请包括胡蝶在内的一众电影明星在大华饭店举行"美亚时装表演",出席者名流如云,轰动一时。1933年,美亚发展到有近千台丝织机,3 000多名雇员,创造了中国丝织工业的记录。

"外祖父为人甚为斯文,我记得他永远是穿着雪白的衬衣,很少见他大声呼喊,但他眉宇之间有一种威严。母亲说很多人见了外公都敬畏他,他不用多说半句话便能让对方信服。"香港溢达集团董事长杨敏德在蔡声白的传记里回忆道。

另一位与小楼有关的"海归"是建筑师奚福泉。1926年,莫觞清买下武康路2号居住,分出一部分花园给女儿女婿。小夫妻俩聘请上海著名建筑师奚福泉设计了4号小楼。

奚福泉是中国第一代建筑师,在德国留学时师从研究中国建筑的名师恩斯特·柏石曼。旅德期间,奚福泉深受德国现代建筑运动的影响,回国后作品多呈现纯粹的现代风格,其中虹桥疗养院是载入上海建筑史的名作。

20世纪50年代,莫家离沪赴港。1978年,蔡声白的女婿杨元龙也进入纺织行业,创办溢达集团,同样经营成功。后来他的女儿杨敏德继承企业,购回了位于武康路2号和4号的两座旧居。古典风格的2号别墅有着偌大的弧形阳台,成为她在上海的住所。摩登精巧的4号则改造为关于几代工业家历史的纪念室,十分贴切。

莫家是个大家族,纪念室一楼用整面墙来展示庞大复杂的家谱树。为了表现家族的纺织业背景,从窗帘、挂毯到灯饰,都别出心裁地用了特色织物装点。

昔日男主人的书房也恢复成民国时的模样。明亮的窗前有一张深色写字台,点着墨绿色台灯,桌上放着蔡氏隽秀的书法,仿佛墨迹未干。

几年前,杨敏德重读外祖父的文章,深深感到"祖父那一代的留学生回国后希望能做到的事,应该是我们这一代要接下去努力完成的。希望我们能薪火相传,能做多少便做多少"。

参观指南

武康路4号作为私人博物馆只对公司客户开放,但其武康路上的外立面充满设计感,值得细细欣赏。最美是月夜,婆娑的树影与或圆或方的窗相映成趣。

蔡声白全家福　Family photo of Tsai

No. 4 Wukang Road is a chic, modern villa. The most striking decoration inside the house is a gigantic family tree that covers an entire wall. The building's fate has been tightly entwined with a legendary family for more than half a century.

"Great-grandfather bought No. 2 on Wukang Road and moved there in 1926. He gave a part of the garden to his eldest daughter, Pearl, and son-in-law, Hsiung Tsai. My grandfather hired Xi Fuquan, a well-known Shanghainese architect, to build No. 4 Wukang Road," wrote Marjorie Yang, chairperson of the Esquel Group in her book *Hsiung Tsai*.

"We were very fortunate to have been able to reacquire both properties and have turned one of them into a small gallery showcasing the lives of generations of industrialists," she adds.

Her great-grandfather Moh Shangqing was an influential silk tycoon in old Shanghai. Moh who owned more than 10 silk factories sits atop the large family tree on the wall. To mitigate risks caused by fluctuating global silk prices, Moh decided to enter the silk-weaving business and founded Mayar Silk Mills in 1917. Later he employed his son-in-law, a graduate from Lehigh University in the US, to manage the company.

It proved to be a wise decision. Tsai adapted American management techniques and achieved great success. By 1930, Mayar had 859 looms and more than 2,000 employees.

As one of China's first modern silk companies, Mayar used modern machines rather than traditional family mills that had been used for thousands of years. The modern way of production avoided handmade flaws and improved product quality. Many old Shanghai movie stars chose to wear Mayar's silk.

The Moh family left Shanghai for Hong Kong in the 1950s. Tsai's son-in-law, Yang Yuan Loong, restarted the textile business, Esquel Group, in 1978. It has also been very successful. His daughter Marjorie Yang inherited the business and used No. 2 Wukang Road as her local residence after reacquiring two villas on the street around 2006.

No. 4 has been renovated and is now a private gallery exhibiting the family's history and their textile business. Graced with flowing curves and large glass windows, this simple-cut, utterly modern villa was designed by a Chinese architect who had studied in Germany.

Architect Xi Fuquan was a student of famous German professor Ernest Boerschmann, who had published six academic studies on Chinese architectural history. He suggested Xi study the Qing Dynasty (1644–1911)'s royal mausoleum for his PhD thesis.

During Xi's time in Germany, the country's modern architectural movement, including Der Ring, hosted many lectures and exhibitions to promote modern architectural concepts. So when Xi returned to China, most of his works was were purely modern, such as Hongqiao Sanatorium, which is regarded as a signature piece of modern architecture in Shanghai.

Esquel Group used a variety of textile products to decorate the house, from curtains, hanging carpet to yarn cone. Even the light is embellished with cloth. They also used cotton from the company's Xinjiang farm to make a stool and bought an antique weaving machine from Jiangsu Province to showcase the origins of the textile industry.

In addition to a rainbow of textiles from antique silk to cotton and cloth, No. 4 also has a recreation of a 1940s study with antique table, original suitcases and calligraphy by Tsai, who had lived in the building.

"My grandfather was ever gentle and rarely raised his voice," Yang recalls. "He dressed as a gentleman and sometimes wore a hat. He favored white shirts and that may be why I have always been partial to clean white shirts. Mother said he commanded the respect of others and often, with the raise of an eyebrow drew fear from his staff.

"When I was with him, I was too young to understand, but his words did leave an impression. Now I appreciate and benefit from his wisdom. I am inspired to carry on his ambitions," she adds.

Tips

The building is open as a private museum and to the company's guests and friends only. The modern façade can be admired from Wukang Road.

WK-12
谭邸的庭院
Tan's Courtyard

　　武康路 12 号是著名建筑师谭垣设计的自宅。这座现代风格的别墅外观"摩登"，有明亮的转角弧形窗和铸铁栏杆阳台，面向武康路的东立面点缀有小圆窗和通高两层的长窗，生动耐看。别墅有一个精巧的庭院。

　　谭垣是一位有传奇色彩的建筑师。1903 年他在上海出生，因为有四分之一美国白人血统，长得很像外国人，出席活动经常被当作外宾。他不仅外貌像"外国人"，而且还颇有"外国脾气"，为人耿直，心直口快，不擅长中国人婉转的说话方式。

　　谭垣毕业于美国宾夕法尼亚大学建筑学专业，1930 年回国后在上海范文照建筑事务所工作，又先后在中央大学和之江大学建筑系任教。1952 年起，谭垣担任同济大学建筑系教授，为中国自主培养第一代建筑师作出贡献。他主持设计的"上海人民英雄纪念碑"和"聂耳纪念园"

方案曾获设计竞赛一等奖,晚年著有《纪念性建筑》一书。

谭垣在上课时用英语讲授建筑术语,从他的课堂里还传出了"谭立面"的名声,因为他特别强调建筑立面设计的重要性。他总是用一支6B铅笔为学生改图,巧妙安排好立面的几个要素,如比例、尺度和韵律。"谭师"的修改常让学生感到豁然开朗,受益匪浅。

谭垣也在武康路的家中收授学生,类似西方名师带徒的工作室教学法"Atelier"。曾在谭家学习建筑设计的朱亚新教授回忆道:"谭宅庭园不大,沿围墙筑有花坛。谭师偏爱叶之绿,不羡花之艳。油亮大叶的龟背竹是他的最爱。室内郁郁葱葱。桌上、地上、墙面上满布各色绿叶,并伴有笼鸟唧唧,生机盎然,充满着中国文人优雅的气息。"

武康路的谭宅有两层,楼下曾是客厅、餐厅和厨房,楼上有三间卧室。1996年谭垣去世后房子易主,墙面曾被租户贴上闪亮的马赛克,幸而最终又恢复了素白的立面。如今一楼是咖啡馆,二楼是家居店,都安静幽雅。

2010年,谭垣的学生们,包括建筑大师张开济和吴良镛等,撰写了

谭垣与同事 Tan and his colleages

近30篇回忆文章,连同谭师的作品集结为《谭垣纪念文集》出版。谭垣之子谭乐在文集里写道,父亲酷爱自然和绿化,鸟雀使他陶醉,最高兴的时刻是和学生们在一起。

他还提及父亲常诙谐地说:"我一生并没有什么积蓄,但是却有着别人没有的财富——我拥有众多的学生,从部长到普通建筑师,从七、八十岁的老专家到刚入校的新同学,遍及全国全世界。"

这本文集里印有一张谭垣与同事坐在台阶上的合影。武康路12号的庭院里,谭师合影的台阶犹在,边上放着一大盆油亮的龟背竹。

参观指南

武康路12号作为咖啡馆对外营业,周一到周五空位较多,周末最好提前预约。天气好的日子可以在谭教授的绿色庭院里喝茶。

No. 12 Wukang Road was the former residence of the famous architect Tan Yuan who designed the house by himself.

The modern-style villa features bright, curved window and a balcony with cast iron railings. The eastern façade on Wukang Road is vividly adorned by long, steep or small round-shaped windows. The villa has a delicate courtyard.

Born in Shanghai in 1903, Tan was a legendary architect who had the appearance of a foreigner since his grandfather was an American. Often mis-regarded as a "foreign guest," he was renowned for a straightforward "foreigner's temple".

An architectural graduate from University of Pennsylvania, he returned to Shanghai in 1930 to work for the firm of Chinese architect Robert Fan and later taught architecture in two Chinese universities. Since 1952 he became a professor of Tongji University's architectural school and contributed in teaching China's first generation of self-trained architects.

He was also an award-winning architect. Two of his designs, Shanghai people hero monument and memorial garden for musician Nie Er, are classic pieces. He also authored a book named *Memorial Architecture*.

The professor was nicknamed "Tan Facade" by his students because he strengthened the importance of designing architectural façade. He always revised students' drawings with a 6B pencil in an inspiring way.

Tan also taught students at his Wukang Road home, which was like an atelier.

"Professor Tan's courtyard is not big, which is laced by flower terraces along the walls. He preferred green plants than flowers. His favorite plant was monstera with big, glossy leaves. The interior of his home was green here and there, graced by a variety of green plants on the table, on the floor or on the walls. Birds in the cage were chirping. This lively home was full of an elegant ambience of Chinese scholars," recalls professor Zhu Yaxin, a former student of this "Atelier."

The two-floor house featured a sitting room, a dining room and a kitchen on the ground floor and three bedrooms upstairs.

Tan's house changed its owner after he passed away in 1996. The walls, which had been covered by shining mosaics by a later user, was restored to pure white. Now the villa houses a quiet, stylish café/boutique shop.

Articles in memory of Tan by his students, including architectural maestros Zhang Kaiji and Wu Liangyong, along with Tan's works were compiled into a book named *Commemorative Accounts of Tan Yuan*.

In the book his son Tan Le wrote about the professor's love for nature and his students.

"I have not much savings, but I have fortunes that other people don't usually have—my students, from ministers to ordinary architects, from elderly experts in their 70s and 80s to freshman students, across China and the world," Tan Le recalled his father's words.

The grey-tuned book published a photo of Tan with his colleagues sitting on the steps in the courtyard. The steps are still there in No. 12's courtyard with a big pot of glossy monstera by the side.

Tips

The building is open to the public as a café. The limited seats are more available from Monday to Friday. I'd suggest have a cup of tea in Tan's courtyard in nice weather day.

WK-40-1
让历史哗然的恬静小楼
A Quiet, Renowned Villa

武康路40弄1号是一座平和淡雅的西班牙式住宅。1938年9月30日，民国第一任内阁总理唐绍仪在此遭暗杀，打破了小楼的恬静。这位著名政治家、昔日清朝留美幼童的人生在武康路戛然而止。

小楼建于1932年，由同样留美归来的中国建筑师董大酉设计。董大酉曾被委以重任，担任上海特别市政府"大上海计划"的顾问。他的工作成果瞩目，今日犹存。位于江湾的原市府大厦、博物馆和图书馆等标志性建筑设计新颖，既满足现代功能，又带有浓郁的中国传统建筑的神韵。

在设计武康路小楼时，董大酉选择了西班牙风格。造价只有英式别墅的一半不到，而装饰效果却很漂亮，因而西班牙风格住宅当时一度盛行。

19世纪70年代，清政府先后派出四批共120名幼童赴美留学，唐绍仪是其中一员。出国前，小唐绍仪与后来担任清政府邮传部副大臣的梁如浩合影一张。老照片上这两个眉清目秀的男孩英气勃发,一派"少年中国"的气象。

由于留学期限长达15年,因而中国父母们对于把儿子送到遥远而陌生的"花旗国"相当不舍。留美幼童李恩富在1886年出版的回忆录里描述了与母亲告别的情景,"我没有拥抱她,也没有亲吻她。哦,这在中国传统礼仪中可不是体面的做法。我所做的就是向我的母亲磕了四个头。她想装出高兴的样子,但我能看见泪水在她眼睛中转……"

留美幼童学成回国后,历经晚清政坛的跌宕起伏,他们有人成为中国矿业、铁路业和电报业的先驱,有人担任了中国最早的大学校长和外交官。唐绍仪在清政府担任要职,后加入孙中山的革命队伍,当选内阁总理,但因为与袁世凯的政治分歧而辞职。

在动荡的1937年,这位命运曲折的离任总理住进了女婿诸昌年位于武康路的家,以赏玩古董为乐,当起了寓公。不过侵华日军占领上海后打起了唐绍仪的主意。他们希望笼络一批有政治影响力的中国人。与此同时蒋介石的重庆国民政府也在做他的工作。

不知为何,这位前总理对双方都态度暧昧。重庆政府担心他最终被日本人拉拢,决心派军统特务刺杀唐绍仪。当年这起暗杀事件被广为报道,震惊全国,就连《北华捷报》都在醒目位置上刊登了一篇配有唐绍仪照片的详细报道:"袭击发生在昨日上午9:30,当时四人乘车而来,车停在麦琪路上(今乌鲁木齐中路),后发现被弃。其中一人穿欧式服装,由两位提着篮子的人陪同……"

报道还提到,来人声称有古玩珍奇展示,敲开门后得以在一楼画室单独面见唐绍仪。短短15分钟后,仆人发现唐绍仪头部为利斧砍伤,送医后不治。

唐绍仪殒命的洋房有着西班牙建筑特有的米色拉毛墙面和褐色筒瓦,几处不多的铸铁花饰精致耐看。但最吸引人的还是主入口的精美装饰,结合螺旋柱与复合柱式,两柱间的券门饰有贝壳图案。

参观指南

武康路40弄是一个安静美丽的院落,可以欣赏弄内1号的外立面。4号是医学教育家颜福庆的故居。

The villa at No.40 Wukang Road is a quiet, beautiful Spanish-style house. However, it's famous not for the Chinese architect who designed it, but for the assassination of Tang Shaoyi, the first premier of the Cabinet of the Republic of China.

Tang was murdered in the house on September 30, 1938.

The building was designed and built in 1932 by renowned Chinese architect Dong Dayou, a graduate of the University of Minnesota. One year before, he had created his masterpiece, the Shanghai Government Administrative Center (now the office building of the Shanghai University of Sports), a signature work in the city's Greater Shanghai Plan.

When designing the Wukang Road villa, Dong chose a Spanish style, likely at the request of its owner, C.N. Chu, who was Tang's son-in-law and a former Chinese envoy to Norway and supervisor at the Shanghai Customs office. Spanish style was very popular for small villas at the time. Compared with the British country-style homes, Spanish designs were cheaper to build but still looked nice.

Born in a tea trader's family near Tong Ka Bay in Guangdong Province, Tang Shaoyi was chosen as one of the 120 students sent by the Qing Dynasty (1644–1911) government to study in the US between 1872 and 1875. According to the book *Chinese Educational Mission Students*, the students included 84 boys from Guangdong Province, seven of them from Tang's hometown.

After returning from the US, Tang worked in several high positions within the Qing government but he later joined Sun Yat-sen's revolution, becoming the first premier of the Cabinet of the Republic of China, a job he later quit over political disagreements with powerful general Yuan Shikai.

The retired premier chose to live in his son-in-law's home on Wukang Road in 1937, where he passed his time collecting antiques.

At that time the Japanese government was eager to bribe former Chinese politicians to be their "political puppets." Their goal was to solidify the areas they occupied after invading China. Meanwhile, Chiang Kai-shek's government, which had retreated to Chongqing, had urged Tang to join them there.

The retired premier took up a neutral position and a Chiang military intelligence agent murdered him to prevent him from working with the Japanese, an incident that shocked locals when newspapers reported it.

"The attack was made at 9:30 o'clock yesterday morning when four men arrived in an automobile which they parked in Route Mag…One of the men, wearing European clothes, accompanied by two others carrying baskets, went to the door and asked admission…The men were admitted to the ground floor and MR. Tang was summoned. When he arrived, he gave orders that he wished to be alone with two of the men in the drawing room. About fifteen minutes later the servants discovered that their master had been murdered and that the men had disappeared," the *North-China Herald* reported on October 5, 1938.

In Shanghai, the butter-hued villa on Wukang Road is graced by Spanish red tiles, small arches and spiral columns.

The villa's most attractive feature is the finely-ornamented main entrance. The arched door is positioned between two spiraling composite columns while the decorations of shells, twists and grass adorn the small space above the door in elegant solemnity.

The villa is also significant as it is a living reminder of the Greater Shanghai Plan, a city initiative to revive "Chinese-style" designs.

This was the city's first true urban blueprint. It was initiated by the Kuomintang government in the late 1920s. Since downtown Shanghai was mostly occupied by foreign concessions at the time, planners looked to a vast area in the city's northeast and selected Jiangwan Town to build a new center.

Architect Dong Dayou's version of Shanghai City Hall is reminiscent of a Chinese imperial palace with large roofs, upturned eaves, huge scarlet wooden gates and exquisitely painted traditional patterns on its exterior. But the general structure, the façade and the entrance clearly showcase the eclectic style popular in the early 20th century.

He also designed the Shanghai Library (now Yangpu District Library) and the former Shanghai Museum (now Changhai Hospital's Screening Building) as part of this ambitious urban plan, which was unfortunately halted after Japan's invasion.

Life can be fleeting, especially for those in powerful positions. But sometimes buildings remain to remind us how history can be cruel and dramatic.

Tips

The building is a private residence and is concealed inside the compound of 40 Wukang Road, a pleasant neighborhood dotted with a rainbow of villas including Building No. 4, the former residence of renowned modern Chinese medical educator Yan Fuqing.

绿树荫浓的颜氏旧居
Dr. Yan's Home

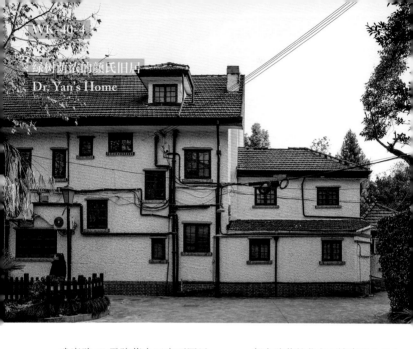

　　武康路40弄院落由四座不同风格的西式别墅围合，宽敞静谧，近年荣获"星级弄堂"的美誉。20世纪40年代，著名医学教育家颜福庆在4号小楼居住。

　　颜氏旧居是武康路一带常见的英式乡村风格住宅。砖木结构的小楼建于1923年，红瓦斜屋顶，姜黄水泥拉毛墙面。装饰不多，但窗框有精致的红砖饰带，与黄色外墙对比鲜明。

　　朝向院落的北立面镶嵌着十几个木窗，大小不一，高低错落，既满足采光需要，又让简洁的立面丰富耐看。

　　20世纪二、三十年代是上海的黄金时期，颜福庆的医学事业也相当丰富精彩。他游说政界和金融界的实权人物，辛苦筹款终于建起集医学院、医院和研究院于一体的医事中心。这是第一个中国人自己的医事中心，将临床实践与教学科研有机结合，达到同期欧美医学院同类水平。

他还说服圣约翰校友、沪上工商名人叶澄衷之子叶子衡捐出自己的花园,开办结核病医院。此外,他创办了中华医学会,沪上英文报刊时常报道关于他的讲话和活动。

1928年11月7日,颜福庆一张精神矍铄的照片登上美国报纸《密勒氏评论》"Who's who in China"栏目,这是专门向西人介绍华人名流的专栏。

"颜福庆博士1882年7月18日出生于上海,1903年毕业于圣约翰大学医学院。大学毕业后他去南非工作,后赴美深造,1909年获得耶鲁大学硕士学位。"报道对他的医学事业高度评价,认为颜福庆"不仅仅是卓越的医生和医学教育家,还做了大量公共卫生和慈善方面的工作。为提升中国医学教育水平和医疗水准,他甘于奉献。"

颜氏旧居现在是一家公司用房,未对公众开放。铁门紧锁,围墙上有高高的竹篱,但仍可以看到故居有偌大的花园。几棵参天大树比武康路上的行道树还要高大茂盛。

据说,江湾长大的颜福庆虽然思维西式,但一生讲一口上海土话。他认为,"医学为民族强弱之根基,人类存在之关键,要为公众利益为目的去学医,而不是赚钱。"

1970年,颜福庆在"文革"中去世,但他培养出了一大批中国医学的优秀人才。他创办的医事中心发展为今天的复旦大学上海医学院和中山医院,叶家花园成为上海市肺科医院;他提出建立的公共卫生防疫系统,将医学的关注点从疾病转向人群和社区,延续至今。

说上海土话的留洋博士辛勤种下的种子,仿佛已长成他武康路旧居前的大树,仍在荫泽和护佑这座城市的人们。

参观指南

"星级弄堂"里的1号和4号都住过中国近代史上有影响力的人物,这个院落值得细细品味。

As Wukang Road's rejuvenation project was nearing completion in the early winter of 2009, the Xuhui district government held a simple ceremony to hang the name plate of "former residence of Yan Fuqing (or F. C. Yen)" on the wall of 40 Wukang Road.

Awarded as a "star-rate lane," the ample, tranquil compound on No. 40 houses four western villas in different architectural styles. Dr. Yan Fuqing, the famous medical educator and founder of both Shanghai Medical College and Zhongshan Hospital had lived in building No. 4 of this compound in the 1940s.

Yan's residence is modeled after a British country villa, a style commonly seen in the Wukang Road neighborhood. Built in 1923, the villa features ginger-hued stucco walls holding up a steeply sloping, red-tile roof. Ornaments are used with great restraint but the window frames are laced with delicate red-brick strips which contrast with the yellow walls.

Facing the yard, the northern façade is mosaicked with more than 10 wooden windows at different heights and sizes, which not only introduce ample sunlight into the house but also enrich the otherwise simple façade.

Back in the golden 1920s to 30s, Dr. Yan also lived such an enriched life

as he lobbied prominent politicians and financiers to establish a medical center comprising a medical school, a hospital and a medical academy all in one institution. This was China's first medical center to organically combine a medical practice with education and research, which reached the standards of its European and American counterparts.

He also persuaded Ye Ziheng, a classmate from St. John's and son of Shanghai tycoon Ye Chengzhong, to donate his garden as a tuberculosis hospital. In addition, he founded the China Medical Association.

On November 7, 1928 he was introduced by the column "Who's who in China" of *The China Weekly Review*.

"Dr. F. C. Yen was born in Shanghai on July 18, 1882 and graduated from St John's Medical School in

1903. After college, he traveled to South Africa while serving as a medical officer in the Chinese Labor Corps from 1903-04. From South Africa he went America for further study and graduated from Yale University with an M.D. cum laude in 1909. From America Dr. Yen went to England where he conducted research at the University of Liverpool and secured the degree of D. T. M." The newspaper sang high praise of Dr. Yan, who was "not only prominent as a doctor and a medical educator, but also worked to promote public health and other philanthropic work. He is a devoted worker in promoting medical education and in raising the standard of medical practice in China."

Yan's residence is now used by a company and not open to the public. The iron gate is always locked but a spacious garden behind is still visible. Several trees in the garden are even taller and more flourishing than avenue trees along Wukang Road.

Having grown up in the Jiangwan area, Dr. Yan had a western mind but always spoke local Shanghai dialect. Throughout his life he held the belief that "medical science is fundamental for a nation's development and humans' living." The purpose for medical study, he always maintained, should be for public interest, not for profit.

Dr. Yan had helped to train a large group of excellent medical practitioners for China before he passed away in 1970. The medical center he founded developed into the Shanghai Medical University and Zhongshan Hospital. The former became the medical school of Fudan University. Ye's garden is now the Shanghai Pulmonary Hospital. The public health and epidemic prevention system proposed by him has continued until today, which turned medical focus from disease to congregation of people and community.

It seems the seeds planted by Dr. Yan had grown to be the unusually tall, flourishing trees fronting his former home on Wukang Road, which are still benefiting and protecting people of this city.

Tips

Two buildings in this "star-rated lane", No. 1 and No. 4 have housed influential figures in modern Chinese history. This is a yard worth wandering and pondering over.

WK-67

低调的陈公馆
A Politician's Home

原武康路菜场对面的一座灰色别墅里，曾经住过国民党高官陈立夫。他长达101年的人生跌宕起伏，尝过权力巅峰的滋味，落魄时靠养鸡为生，晚年东山再起。

20世纪40年代，陈立夫将上海公馆设在武康路67号，而他的哥哥陈果夫就住在武康路另一头，也是一座低调的灰色小楼里。

陈氏兄弟跻身于民国时期蒋、宋、孔、陈"四大家族"，这四个家族控制了20世纪上半叶中国的政治和经济命脉。

陈立夫出生于浙江吴兴，1926年留美归来，本来打算学以致用，当一名采矿工程师。然而命运蹊跷，经哥哥陈果夫劝说，他担任了蒋介石的秘书，开始走仕途人生。到20世纪30年代，这对陈氏兄弟掌握了国民党的党务大权和中统特务机构，一度权倾朝野，建立了政治派系"CC系"。

徐汇区文化遗产档案记载，陈

立夫的别墅占地面积663平方米，是一栋三层楼的砖木结构建筑。外墙以水泥粉刷，特色是一楼有一间六角形房间。如今，几户人家共同居住在这座老房子里。室内光线昏暗，满是尘埃和杂物，但绛红楼梯上的雕花图案仍旧精致耐看。研究上海旧法租界的同济大学副教授刘刚认为，与南京的民国公馆建筑相比，陈立夫的武康路别墅富有上海特色，风格简洁、自由而活泼。

1950年，陈立夫50岁，正好是他百岁人生的一半，也是人生的转折点。此时陈氏兄弟和"CC系"成为国民党失守大陆的替罪羊，被排挤出权力中心，陈立夫被迫带着家人远走他乡，到美国新泽西靠经营养鸡场卖鸡蛋为生。2001年，陈立夫的三子陈泽宠回忆刚去世的父亲："我的父亲总是喜欢用他的智慧去解决问题并帮助他人。我们觉得他有能力处理任何难题。当时我只有11岁，而父亲已年迈，我们要在新泽西搬运鸡饲料，每袋重约100磅。父亲便想办法设计并制作了一个木制机器，用来运输和倾倒饲料。此后，其他的一些美国农场也广泛使用这项技术。"

留美的18年间，陈立夫还卖过一些自制传统食品，如做月饼用的咸蛋黄，而"陈立夫辣椒酱"在美国华人圈里很有名气。

因为与蒋经国的交情，1969年陈立夫应邀返台，提出"以中国文化统一中国"的方案，在岛内引发震撼。他本人也成为中国传统文化的守护者，在担任台湾中国医药学院董事长的30年间，出版了几十部专著，推进海峡两岸中医药事业的发展和交流。

陈立夫的人生让人联想到武康路的另一位居民——著名作家巴金。两位老人都活了101岁，人到中年都遭遇重大挫折，晚年却神奇地开创了一番新局面。

2000年，陈立夫撰写了一篇题为"我怎么会活到一百岁"的文章，将自己的养生之道归结为："我能熟睡，不发脾气，记忆力强，饮食有节，多食果菜，水沸始饮。"

参观指南

建筑是私人住宅，不对外开放，但可以漫步小小的院落，欣赏独特六角形房间的外立面。建议事先读点陈氏兄弟的历史故事，再探访武康路107号的陈果夫故居。

The gray-walled villa on 67 Wukang Road was once the home of Chen Lifu who had been an influential Kuomintang official, a farmer and a traditional Chinese medicine promoter over an eventful life that lasted 101 years.

The upper-class residential area in the western area of the former French Concession, including the neighborhood of Wukang Road, was chosen by Kuomintang military and governmental officials as their homes after they took over Shanghai following China's victory in War of Resistance against Japanese Aggression in 1945. Chen's older brother Chen Guofu lived in a villa on the other end of Wukang Road.

The Chen brothers were one of the four "big families"–Chiang, Song, Kung and Chen–of the Republic of China. These families had controlled much of the country's finances and politics.

Chiang Kai-shek controlled the military power but assigned financial matters to T.V. Song and H. H. Kung, and Kuomintang party affairs to the Chen brothers.

Born in Wuxing, Zhejiang Province, Chen Lifu had dreamed of being an engineer to help the country's modernization efforts after receiving his master's degree in mining engineering at the University of Pittsburgh in Pennsylvania, United States.

But at the advice of his older brother, Chen Guofu, who had already worked with Chiang Kai-shek, the young man became Chiang's secretary in 1926.

In the 1930s, the Chen brothers controlled the Organization Department of the Kuomintang government and Chen Lifu headed the department's investigative section, a renowned group of special agents. As a result of the Chens' significant power and influence within the government, they formed a political faction known as the "CC Clique."

According to Xuhui District Record of Cultural Relics, Chen's gray villa was built in the 1940s. Covering an area of 663 square meters, it remains a three-story, simple, brick-and-wood villa. The façade is covered with cement. The corner of the ground floor features a protruding hexagonal room. The mini garden fronting the villa was at some point paved with cement.

Several families now share the villa. The dim, dusty interior endows the home with a feeling of history.

Tongji University associate professor Liu Gang has observed that Chen's villa contains many "Shanghainese characteristics," which have a freer style and are comparatively vivid compared with the congregation of official mansions in Nanjing, the former capital of the Kuomintang government.

Chen Lifu's political career ended after the Kuomintang retreated to Taiwan after 1949. The Chen brothers and their CC Clique were blamed for Chiang's failure against the Communist Party of China and were pushed out of the inner core of political power in Taiwan.

In 1950, when he was 50 years old, Chen took his family and immigrated to New Jersey, US where he settled down and ran a chicken farm, making a living by selling eggs.

In the autobiography, he wrote: "I had served Chiang and many others, but eventually I was not well understood by them. From now on, I will never serve anybody and will never have this kind of feeling again when looking after the chickens."

In 2001, Chen's third son, Chen Zechong, wrote an article about his father for Taiwan's Central Daily after his dad's death.

"My father always loved to use his brain to find solutions and help

others," he wrote. "We think he was capable of solving any problem. I was only 11 years old, but my father was already pretty old when we had to carry the chicken feed in New Jersey. Each bag weighed about 100 pounds. He then racked his brain to design and make a wooden machine to transmit and dump the feed. It was widely used by other American farms afterward."

During his 18 years in the US, Chen also made a living by producing traditional Chinese foods ranging from pickled duck egg yolks, used as a stuffing in moon cakes, zongzi (traditional Chinese food made of glutinous rice wrapped in leaves) and chili sauce. His homemade Chen Lifu Chili Sauce was popular among Chinese Americans.

Owing to his long-term good relationship with Chiang's son, Chen was invited to return to Taiwan in 1969. He then spent his later years as a "guardian of traditional Chinese culture."

During the last 30 years of his life, he served as director on several committees for reviving traditional Chinese culture and TCM in Taiwan. He published 30 books in this field and proposed a famous idea to culturally unite China's mainland and Taiwan.

In an article titled "How I lived to be 100 years old" in 2000, he attributed his longevity to the ability to fall asleep quickly, a good temper, good memory, a healthy diet with mostly vegetables and drinking only boiled water.

"I never liked politics, which often requires mean methods in the struggle for power and profits. If not for Chiang's decision, I would have been a mining engineer long ago... To nurture your heart you need peace and first of all, a clean ideal and simple life," wrote this 100-year-old man who had so much power in his hands when he had a home on Wukang Road.

Tips

I suggest visitors also make their way to Chen Lifu's brother, Chen Guofu's residence at No. 107.

WK-99

风笛悠扬的大班洋房
Pipe Music from Taipan's House

1931年6月1日一早，福开森路99号英商正广和大班家里传出悠扬的乐声。大班麦克格里格为隆重招待万国商团的苏格兰连队，请了一支名叫"绿色霍华德"的乐队到家中演奏，而苏格兰连队的风笛手也加入了演出。

万国商团简称"S.V.C."，是上海租界当局以保护侨民为名成立的一支准军事化武装。这支商团名副其实的"万国"，沪上侨民人数较多的国家纷纷在商团中建立连队，如德国连队、美国连队、意大利连队、苏格兰连队等。

麦克格里格大班也是上海苏格兰社团——圣安德鲁斯协会的主席，他在招待会上发表演说，强调商团保护租界和侨民的重要性，呼吁苏格兰年轻人投入此项工作。

对上海大班们来说，租界的稳定安全是商业利益的保证。就在风笛招待会举办的数月前，这位大班亲自在位于福州路的正广和总部主持了年度股东大会，骄傲地向股东们宣布利润增长可观。

正广和洋行源于英商创办于1864年的广和洋行，起初从事洋酒和啤酒业务，1882年更名"正广和"，取意"正本清源，广泛流通，和颜悦色。"1892年，正广和汽水厂成立，以"Aquarius"（水瓶星座）注册为英文商标，生意兴隆，远销海外。

1933年，英文《大陆报》刊登报道介绍正广和汽水既美味，又品质纯正。汽水品种多样，最受欢迎的有香蕉、柠檬、鲜橙等口味，而正广和干姜啤酒的质量世界一流。

报道还刊登了一张正广和公司的照片，与大班所住的99号英国乡村别墅非常相似，露明木结构的建筑有坡屋顶和高耸的烟囱。

福州路正广和公司
Caldbeck, Macgregor & Co. Ltd, Fuzhou Road

而1954年公私合营的正广和公司，在20世纪90年代遭遇了与洋可乐的激烈竞争，一度低迷，后来又因市民饮用水工程东山再起。

2014年，正广和举行活动庆祝品牌的150岁生日，宣布根据消费者的需求，推出一款天然无气苏打水的新产品，品牌就叫"Aquarius"。

参观指南

别墅虽然不对外开放，但透过铁门可以欣赏建筑高耸的坡屋顶、老虎窗、别致的烟囱和齿形装饰。

On the morning of June 1st, 1931, Orchestral pipe music could be heard coming from the residence of J.F. Macgregor at No. 99 Ferguson Street.

As president of the St. Andrew's Society, the chairman of Messrs. Caldbeck Macgregor & Co., Ltd, was entertaining members of the Shanghai Scottish, S.V.C. at his residence. The Green Howards played alongside the pipers of the Shanghai Scottish Band under the direction of Pipe-Major Welstead, according to the report of the *North-China Herald* on June 3.

S.V.C., or The Shanghai Volunteer Corps was a multi-national, mostly volunteer force controlled by the Shanghai Municipal Council.

During the reception, Mr. Macgregor explained that he felt special pleasure in welcoming the Shanghai Scottish Band because he "desired to impress on the young Scottish men of Shanghai the importance of doing their part in the defense of the Settlement."

For Shanghai merchants like Mr. Macgregor, commercial interests relied on the safety and stability of the settlement. Two months prior to the pipe reception, while presiding over the annual shareholders' meeting in the company office on Fuzhou Road he proudly announced a considerable increase of profits during the past year.

The history of his company dates to

正广和汽水　Original Aquarius' soda

1864, when George Smith & Co. ventured into the Wines and Spirits business in Shanghai. John Macgregor and E. J. Caldbeck took over the company in 1882, changing its name to Caldbeck, Macgregor & Co. Ltd in 1883.

The company's three-character Chinese name "Zheng Guang He" roughly means "purified, widely-loved and pleasing". In 1893 the company launched an aerated drink sub-company, Aquarius, a name referring to the ancient Greek myth of a deity pouring water from a vessel.

The company later emerged as the leading wine & spirits business in the Far East. In 1937, *The China Press* stated that the waters supplied by the Aquarius Company "are not only delicious and varied, but are guaranteed to be absolutely pure." The company offered a wide variety of sweet beverages. Among the popular fruits flavors were banana-squash, lemonade, lemon crush and orange crush.

The article also published a photo of the company office on Fuzhou Road, which resembled Macgregor's British country villa on 99 Route Ferguson in many ways. Both buildings feature half-timber structure, sloping roof and a steep chimney.

The villa had a new resident after the Cultural Revolution (1966–1976) when renowned entrepreneur Liu Jingji (1902–1997), who served as Vice Chairman of the CPPCC National Committee, moved in. Having worked hard to save the country through industry and business in Shanghai, he founded Shanghai Aijian Group Company, currently listed on the Shanghai Stock Exchange.

The Caldbeck, MacGregor & Co. Ltd became a state-owned company in 1954 and its "Zheng Guang He" brand kept the city's market dominance until the re-entry of international drinking brands into China almost three decades later. To win back the market after an increase in international competition, the company responded to

正广和饮用水　Aquarius' Natural soda

complaints about the quality of tap water in Shanghai and became one of China's first providers of drinking water in large plastic containers. The new service with home delivery made the company popular once again.

In 2014 the company celebrated its 150th anniversary by launching a popular new product, a natural soda water named "Aquarius."

Tips

While the villa is not open to the public, the building's sloping roof, dormer windows, stylish chimney and tooth-shaped ornaments can be appreciated through the iron gate.

巴金故居的笑与泪
Tears and Joy in Ba Jin's Residence

著名作家巴金在武康路住了近半个世纪,他位于113号的故居已对外开放多年,但很少有人知道巴金其实只是这座洋房的租客。

巴金故居常务副馆长周立民在整理巴老的物品时,发现了三张20世纪50年代的房租收据,巴金通过代理人向英国房主支付每月144.2元的租金。周馆长提到,巴金是当时国内唯一不领工资的作家,他用版税来支付不菲的房租。

这座美丽的房子里原来住的是英国药剂师理查森医生夫妇和两个女儿。1924年,一个女儿嫁给了英国帝国化工总裁、上海英国侨民协会主席科乐先生。

科乐喜欢到外滩上海总会打保龄球,热爱划船运动,一家人过着沪上侨民标准的舒适生活。然而,1941年,"太平洋战争"爆发后,外国侨民命运突变。科乐说服太太带着一双幼女回国避难,自己勇敢地选择留在上海坚守公司。1942年,他和很多武康路邻居一起被关进日军位于海防路的集中营。

在集中营,机智乐观的科乐成为300多位被囚侨民的领袖。他运用"外交手段"巧妙与日军周旋,帮助和保护了很多人。

1945年科乐获释,作为心理释放,他开始用文字记录过去三年多集中营艰苦压抑的生活。他并未计划出版这部珍贵的战争回忆录,直到1985年离世前夕,才经友人建议出版成书,名叫《上海战俘》。书中收录了数张科乐在上海的旧照。无论是西装笔挺携娇妻站在花园洋房前,还是衣衫破旧和日本军人在集中营里,这位英国绅士的嘴角始终漾着幽默淡然的浅笑。

科乐获释10年后,巴金一家入住113号。新家占地1400平方米,有三层高的主楼,两座辅楼和偌大的花园,极大地改善了巴金的写作环境。他对新家非常满意,添置了家具,在底层客厅会友,到楼上书房写作。巴金撰文描绘了新家的生活,写道:"全身充满幸福的感觉"。

幸福的日子过了十几年,1972年巴金妻子萧珊因肝癌去世,给他带来很大的打击。但此后,他迎来创作的新高度,写出《随想录》等具有深远影响的巨作。

2005年巴金去世后,他的家人搬离武康路,留下大部分家具和巴金个人物品作为故居陈列。

周馆长还发现,巴老不喜欢丢东西,从50年代开始,他穿过的每一套衣服都保留着。此外还有十几个收音机、大量读者来信和近八万册藏书。

2011年故居开放至今已接待了几十万人。在这个家里的每个角落,从家具到园中草木,仔细看和体会,点点滴滴都是这位百岁老人跌宕起伏、悲欢离合的文学人生。

前面讲到的科乐先生在《上海战俘》里展示了一张自制的圣诞卡,是1943年他从海防路集中营里寄出的。在这张卡片上,科乐先生手绘了自己的集中营生活,曾经养尊处优的他要自己打扫卫生、做饭、洗衣,也有简单的医疗和体育活动。他还手写了一首英文诗,意为:"要向上看,向前看,时间终会疗愈我们……"

参观指南

房子里的一景一物与巴金居住时基本一致,装饰品都是原物。故居开放时间为周二至周日上午10点到下午4点。

Renowned Chinese writer Ba Jin (1904–2005) lived half of his life in a gray-toned garden villa at 113 Wukang Road built in 1923. But it's rarely known that he was only a renter of the big house.

While sorting out the writer's collection of books and documents in the house before opening to the public, Zhou Limin, deputy director of the Former Residence of Ba Jin found three rental receipts from 1955 and 1956.

The receipts revealed the original owner of the house was Maud Pauline Hay, who had commissioned an agent named A. Springborg to sign the rental contract with Ba. The monthly rent was 144.2 yuan (US$23.3), a steep price compared with the average monthly salary of 38 yuan to 68 yuan at the time. Since Ba received no salary from any institutions, he paid the expensive rent by payments for his books.

Covering an area of 1,400 square meters, the Wukang Road villa comprises a three-story main building, two auxiliary buildings and a 400-square-meter garden. Living here seemed to have a positive impact on writing as Ba Jin wrote many famous articles in this house.

Decades before the writer moved in, it was home to Dr. Alan Richardson, his wife and two daughters. One of the daughters, "Bunny", married Hugh Collar (1900–1985), Director of Imperial Chemical Industries (ICI) in Shanghai and Chairman of the British Residents' Association from 1940. The newlyweds lived in the villa on Route de Ferguson for a few years. Maud Pauline Hay was the other daughter.

Collar had lived a comfortable life before the Pacific War changed the fates of Shanghai expatriates. He chose to stay with the company but persuaded his wife to leave Shanghai for good with their two daughters. In 1942, he was put into a Japanese camp on today's Haifang Road.

Again, this man of "courage and invaluable sense of humor" became chief representative of some 360 internees. He had to fight a battle of wits against the Japanese authorities and was known to have helped and save many people.

He wrote a firsthand account during the period of an indefinite leave he was given by his Company after the Japanese surrender, which was published into a book, *Captive in Shanghai*.

Collar seemed have always wore a humorous smile in photos no matter if he was attired in a handsome suit standing with his wife before a garden villa or in a ragged garment juxtaposed against several Japanese officers in the camp.

科乐在集中营绘制的圣诞卡片
A Christmas card made at the camp by Collar, 1943

Collar later wrote that "Four years of daily contact with the Japanese had left me somewhat frazzled. Not quite a nervous wreck, but definitely not easy to live with. This was why I decided to try to write it out of my system. I can well remember the sigh of relief with which I wrote the last few words on a storm tossed freighter on the way back to Shanghai. The exercise worked."

Ten years after Collar was set free, Ba Jin's family moved into No. 113. The writer was so happy that he bought some new furniture, hosted guests on the ground floor and worked upstairs. In an article he wrote about his life in the new home — "I felt overwhelmed by a happy feeling"

His family's happy life lasted more than a decade until the "cultural revolution" (1966–1976) broke out and his wife Xiao Shan died in 1972 of cancer. Her death had a profound impact on Ba, who kept Xiao's urn on the cabinet beside his bed.

After the "cultural revolution," Ba was elected chairman of the Chinese

科乐在海防路集中营
Hugh Collar in the grounds of the Haiphong Road Camp

二战前的上海，科乐在苏州河划船
Rowing on Suzhou Creek in pre-war Shanghai. Hugh Collar is the Stroke

Writers' Association and created the most significant works of his later years including *Random Thoughts*, a painfully honest reflection on the turbulent 10 years.

After Ba died in 2005, his family moved out of the house but left most of the furniture and his personal belongings so they could be displayed to the public.

Director Zhou has found that the writer was a careful man who kept everything including his clothes from the 1950s until his death, more than 10 radios, almost every reader's letter and nearly 80,000 books.

The villa opened to the public as a Ba Jin memorial museum in December 2011 and has received more than 300,000 visitors.

In the book *Captive in Shanghai*, Mr. Collar, the son-in-law of Ba Jin's landlord, showed a self-made Christmas card he sent from the Japanese camp in 1943 while in captivity.

He painted funny cartoons on the card to describe his life in the camp where he cooked, cleaned and washed clothes. Collar also wrote W. E. Henley's poem beside the cartoons — "Upward and forward. Time will restore us."

Tips

The house remains largely the same as when Ba Jin lived here, including even the small vase and decorations on the cabinet. (Open from 10am to 4pm, Tuesday to Sunday).

FXW-147

柯灵的精巧之家
A Tiny, Smart Home

作家柯灵住得离好友巴金家不远，虽然房子小了很多，但正如美国作家芭芭拉·弗拉纳根的一本书名，是个"精巧之家"。

柯灵故居建于1933年，是一座西班牙风格的公寓住宅，由留德归来的建筑师奚福泉设计。米黄色拉毛墙面，外观比例均衡优美。西班牙风格的细部精美，如半圆形券门、螺旋柱券窗、铁艺装饰和红色筒瓦。

1959年12月，柯灵和夫人搬到这里，他在203室寓居的41年里撰写了大量散文、杂文、小说和剧本。

2016年故居修缮后对外开放，与巴金故居、张乐平故居一起，成为武康路一带为数不多的对公众开放的名人故居。

故居一层布置了作家生平展，以及一个书信展厅，展示这位作家的"朋友圈聊天记录"。与柯灵书信往来的大名鼎鼎的文人朋友包括钱锺书、夏衍、梁实秋、傅雷和张爱玲等。张爱玲还是文坛新人时，担任《万象》主编的柯灵慧眼识珠，发表了她的《心经》，成为发现张爱玲的伯乐。

读完文人书信，可以顺着一个按原样修复、直径仅1.2米的螺旋形仆人楼梯，到二楼参观作家的厨房、餐厅、客厅、书房和卧室。

与巴金故居相比，小而精的柯灵故居里有木碗橱、竹篮、花雕酒、布罩台灯、《二十四史》书柜，海派文人的生活气息似乎更加鲜活浓郁。

最别致的家具是卧室的多功能木橱，隔板放下就变为一张小书桌，桌前放着旧藤椅。"文革"期间，柯灵的书房和客厅被查封，卧室内的阳台配上木橱书桌就成为迷你书房。不知为什么，比起家具满当、有点透不过气的书房，这个阳台上的书房空间更亲切宜人。阳光好的日子，坐在窗前创作感觉应该不错。

作家柯灵照片　Writer Ke Ling's photos

故居小小的花园也得到清理整修，角落里低调地放着一座柯灵雕像，但让人印象更深的却是入口处的一张照片。白发作家披了件黑色大衣，好像坐在武康路的一地落叶上，表情严肃而平静。照片对面的墙上写着他的名言，"生活是一部永远读不完的大书。生而有涯，每个人只能读到有限的章节，因此必须认真地读。"

参观指南

故居开放时间为每周二至周日上午9点到下午4点。可以留意这套作家公寓精巧、紧凑、合理的设计，是旧法租界西班牙风格住宅的典范。

Playwright/writer Ke Ling's home on Fuxing Road is only a few minutes' walk from his friend, writer Ba Jin's residence on Wukang Road. Though much smaller in size, the villa mirrors the name of a book by Elle Décor editor Barbara Flanagan — "Smart home."

Ke's home was designed by Chinese architect Xi Fuquan, who also designed the modern villa on 4, Wukang Road. Xi studied in Germany and was heavily influenced by the country's modern architectural movement. When he returned to China, most of his works were pure modern, such as Hongqiao Sanatorium, regarded as one of the city's signature buildings.

For this lovely yellow villa built in 1933, he found inspiration from Spain. Spanish-style villas were trendy at the time as they had a pleasing decorative look and cost less compared with British country homes.

Covering an area of 192 square meters, Ke's former residence occupied the ground two floors and a 150-square-meter garden.

The façade shows a balanced proportion with delicate architectural details for decoration, such as the refined semi-circular arched door, arched windows graced by spiral columns and elegant cast iron embellishments.

The butter-hued cement stucco walls, the wave-shaped decoration underneath the eaves and red tiles are typical Spanish features.

The playwright and his wife moved into room 203 in 1959 where they wrote many articles, novels and plays in the following 41 years.

The villa opened to the public in 2016 after a renovation. Today, the house and the former home of cartoonist Zhang Leping on Wuyuan Road form a "cultural triangle" with Ba Jin's former residence. An area known as the Hunan Community including Wukang, Fuxing and Wuyuan roads once boasted the residences of more than 100 celebrities.

The ground floor features an exhibition of Ke Ling's life and letters from his friends who are all renowned Chinese writers including Qian Zhongshu, Liang Shiqiu and Eileen Chang. It was Ke Ling who first discovered Chang's talent and had her story published while he was serving as editor-in-chief of a popular magazine named *Wan Xiang*.

A tiny, spiral staircase leads to the second floor displaying Ke Ling's bedroom, sitting room, study and kitchen, all restored to their original looks. The layout is concise and smartly arranged. The interior décor is based on the original furniture.

A multi-functional cabinet on the balcony can be switched to a desk. During the Cultural Revolution (1966–1976) when the playwright's study and sitting room were occupied, he used the tiny desk and an old bamboo chair as a mini study that would have been especially nice during sunny days.

The garden contains the original and well-preserved magnolia tree, Japan allspice and oleander along with Ke Ling's statue on the lawn.

One of his photos on the entrance is especially impressive. The white-haired writer wears a black coat while sitting on what appears to be Wukang Road, atop a spread of fallen leaves. His expression is serious and calm. Opposite this photo are his words written on the wall — "Life is a big book you can never finish reading. Life is limited and everyone can only read limited chapters. So we must carefully read this book."

Tips

The residence is open from 9 am to 4 pm every Tuesday to Sunday. Please admire the tiny, concise and smart layout of this villa, a typical example of Spanish villa in the former French Concession.

FX.W-193

英国漫步专家的乡村别墅
The Country Villa of a Country Walker

复兴西路武康路有一座静谧的英式乡村别墅,住过一位著名的英国侨民,他的上海人生过得异常精彩。

这位威尔金森先生曾是英国汤姆逊会计师事务所(Thomson & Co.)的高管。1939年1月,《北华捷报》报道他退休的新闻写道:"威尔金森是那种闲不下来的人,在工作之外总是准备做很多的事情。他是法租界工董局的英侨代表、上海戏剧爱好者俱乐部的支持者、皇家圣乔治社团会员、雷士德基金会原主席……"报道还提到,许多读者对威尔金森在《北华捷报》连载的"乡村日记"专栏印象深刻。"除了度假和出差,他很少中断这个引人入胜的专栏。"

威尔金森是个特别热爱大自然的人,爱好研究鸟类、花卉和昆虫。他把爱好做到了极致,徒步将上海郊区走了个遍,并在英文报纸上连载专栏,介绍上海周边的自然之美。

1929年,他将专栏结集出版为

《上海鸟类》，并请英国艺术家H.格伦沃尔德为专著绘制了大量彩色鸟类插图。出版前他专门在外滩亚洲文会大楼（今外滩美术馆）举办了一场引人入胜的鸟类讲座，不仅播放鸟的叫声，还分享如何用食物和水在自家花园里吸引鸟儿飞来，以便研究它们的习性。

1932年，他又出版了《漫步上海郊区》，分享漫步上海乡间和郊区的路线与见闻。书籍封面由老上海白俄漫画家萨帕乔绘制：威尔金森先生叼着烟斗，惬意漫步，远处的风景有农田、挑担农夫和宝塔。这本书不仅刊登了很多游览路线，还有他悉心撰写的漫步上海注意事项，细致到"建议带苹果解渴"和"穿轻便毛衣"等细节。他还分享了把手表给小孩玩以跟中国人套近乎的妙招。小孩咯咯一笑，他妈妈一高兴准会端上一碗热茶，屡试不爽。

如今，这里是上海市房地产科学研究院（简称"房科院"）的办公楼之一。别墅朝向大花园，三层楼高，有巨大的斜屋顶、米黄色拉毛墙面、生动活泼的连续券窗和拱门。

数年前，在对这座历史建筑大修前，房科院做了一次专业的3D勘测。修缮保护了别墅内的黑白马赛克地砖、深色木楼梯和屋顶的露明木结构。窗上损坏的铜把手和镶木地板也按照原样修复。房科院还特别对建筑进行了防白蚁处理。武康路一带的老洋房深受白蚁的侵扰和损害。

修缮后的别墅如威尔金森居住时那样静谧美丽，每一层都有三个形态大小不同的房间，有的饰有石壁炉，有的有巨大的拱窗，窗外偌大绿色的花园好像一幅美丽的画。

80多年前在这个花园里，有一位英国绅士饶有兴趣地观察来访的喜鹊和夜莺，看完又回到房里写乡村日记和自然之美。这位英国漫步专家可能没有想到，很多年后他家旁边的武康路，也成为人们喜爱漫步徜徉的地方。

参观指南

建筑不对外开放，但在复兴西路上可以欣赏建筑的大部分立面，细心体会威尔金森先生追寻的上海自然之美，以及武康路"城市山林"的氛围。

A British country villa on Fuxing Road near Wukang Road was once home of a renowned British gentleman who lived a full life in Shanghai.

According to the *China Hong List* in 1932, the house at 193 Route de Boissezon (today's Fuxing Road W.) was occupied by E.S. Wilkinson, who worked for Thomson & Co., a chartered accountants' agency from U.K.

The announcement of Wilkinson's retirement in the *North-China Herald* showed how busy his Shanghai life had been. He was president of the local branch of the Royal Society of St. George, secretary of the Lester Trust, a British representative on the council of the French Concession and supporter of the Amateur Drama Club of Shanghai.

A passionate naturalist, Wilkinson researched birds, flowers and insects in his spare time as he walked throughout the city's countryside and contributed a charming series of nature notes under the heading "A Country Diary" in the *North-China Herald*.

In 1929 his columns were edited into a book entitled *Shanghai Birds* with illustrations by British artist H. Gron-

vold. Before launching the book, he hosted an interesting lecture on birds in the Royal Asiatic Society's Hall which is today's Rockbund Museum.

"Mr. Wilkinson has studied birds and their habits not merely from a scientific viewpoint but for pleasure and aesthetic appreciation. He has succeeded in coaxing birds to come to his garden by knowing their habits," the *North-China Herald* reported on December 21, 1929.

In the lecture Wilkinson also shared that "a bird table and a bird bath are valuable lures for those who love feathered creatures. Trees and small shrubs which will provide hiding places for nervous birds and also a supply of insects are also very good inducements."

In 1932 he published another book named *Shanghai Country Walks* about "good walking around Shanghai, especially in the Western country." He offered his reader the choice of an hour's stroll, an afternoon's walk, or a Saturday-Sunday tramp to "where the rainbow ends in the Hills".

The book cover is graced by illustrator Sapajou's drawing of Wilkinson enjoying a walk within the countryside and alongside a Chinese farmer and an ancient pagoda. Wilkinson also shared travel tips such as bringing an apple, wearing light sweater and

威尔金森　E. S. Wilkinson

pleasing a Chinese mother by allowing her child to play with his watch.

After 1949 the house was used as a military medical research organization until it was taken over by the Shanghai Real Estate Science Research Institute in 1975, an organization specializing in studying the phenomenon, problems and technology of the city's real estate market. It's also known for its expertise in renovating historical buildings.

Having been partly destroyed by humidity and termites, the Villa's wooden structure underwent a careful and scientific renovation guided by experts from the institute.

The restored villa was as tranquil and beautiful as when Wilkinson had lived there. The south facade facing the garden features a large double-peak sloped roof topped with red tiles, light yellow rough-finished walls and a vivid line of continuous arched windows and arched doors.

From the north facade near Fuxing Road, steep red chimneys reach out to the sky. The red bricks on the northern walls form a strange protruding pattern, like a mythical sign.

The interior has an antique and stylish look. The three-story villa has three big rooms on each floor, all in different sizes and shapes. Some of the rectangular or polygon-shaped rooms are adorned with carved stone fireplaces or large arched windows, through which the expansive, lush garden might easily be mistaken for a breathtakingly beautiful painting.

It was in this garden more than 80 years ago that a British gentleman spent his afternoons observing magpies and nightingales with great in-

《上海鸟类》 Cover of *Shanghai Birds*

terest before returning inside to write his country diaries and about the beauty of the nature.

This expert of country walks would have no way of knowing that so many years later, these same areas near his home at Wukang Road would be so popular among walkers and nature lovers still.

Tips

The building is not open to the public but much of the façade can be appreciated from Fuxing Road W. Please enjoy the beauty of nature which Wilkinson has pursued and the garden city atmosphere along Wukang Road.

密丹公寓小巧精致，造型特别，又位于街角，在武康路星罗棋布的洋房中引人注目。

简洁的公寓点缀着几何形浮雕和阶梯形装饰。仰头细看，可以发现北面檐口的云朵图案，仿佛正轻轻掠过水泥拉毛墙面，为灰扑扑的公寓增添了灵动的气息。

与外观相比，公寓的楼梯设计更为精巧。设计师发挥"螺蛳壳里做道场"的精神，最大限度地利用了这块六角形的小基地。楼梯仅采用两种朴素的材质——灰色水泥和黄色水磨石，看似无奇，但如果走到顶楼向下看，一圈圈楼梯恰好绕成美丽的漩涡，极富设计感。

长期以来，密丹公寓疑似为法国赉安洋行的作品。作为旧法租界最主流的建筑设计公司，赉安洋行由三位酷帅的法国设计师合伙经营。他们三人的名字——"Leonard""Veysseyre"和"Kruze"组成了洋行的英文名。

赉安洋行是一所既高产又注重设计品质的事务所，留下了包括法国球场总会（今花园饭店）和培文公寓（今上海市妇女用品商店）在内的一系列经典建筑，还有许多类似密丹公寓的小型高级公寓，如复兴路上的卫乐精舍和麦琪公寓。

老上海外文报纸时常刊登有关赉安洋行的报道。1934年7月14日，法文《上海日报》用了一个整版，图文并茂地介绍了60多件赉安作品，令人惊叹。但这些作品里并没有密丹公寓的身影。而根据对建筑原始图纸的最新研究，同济大学郑时龄院士认为，密丹公寓是由比利时义品放款银行建筑部设计的。

1931年，《大陆报》头版的一则图片新闻证实了他的研究。报道还称，由一位中国业主投资兴建的密丹公寓高达四层，占地面积是全市最小的，有两房和四房两种房型。间接照明和现代风的室内装修让这座异国风格的建筑增色不少，是"上海滩独特的公寓房"。

有趣的是，1935年《字林报行名簿》显示，赉安洋行合伙人之一克鲁兹成为密丹公寓顶楼的一名租客。那应该是大楼景观最好的一套公寓，客厅呈六角形，转角的大玻璃窗正好眺望武康路的街景。

细细研究20世纪30年代的几本英文黄页，可以发现公寓昔日的住客几乎清一色都是在跨国公司上

班的外籍人士,他们任职的公司大多都位于外滩一带。看来在幽静的武康路居住,到繁华的外滩上班,曾经是"魔都"外国金领的生活工作模式。

顺便提一下,后来赉安洋行的三位法国帅哥命运迥异。在密丹公寓顶楼小住的克鲁兹离开上海后再也没有消息。凡赛耶后来到越南发展得不错。幸运的他和邬达克一样,大量档案由家人悉心保存,近年来外国学者研究出版了他的传记。

而赉安洋行最有才华的灵魂人物赉安却在20世纪40年代的上海神秘失踪了。老上海英文报纸上一篇篇报道勾勒出这位法国建筑师的"魔都"人生轨迹。他历经事业巅峰、中年丧妻、再婚,太平洋战争爆发后再也没有任何消息。据说,他远在法国的外孙仍在寻找他的下落。

参观指南

欣赏这座建筑简洁立面上的丰富细节,是一件有趣的事。

The Midget Apartments certainly catch the eye when walking along Wukang Road. Shortly after its completion in December of 1931, *The China Press* published a photo of the stylish building with the headline — "Shanghai's Most Unique Apartment House."

The edifice is treated in a simple way, decorated with a geological relief and architectural details shaped like stairs on the corner of the second floor. The cloud-shaped architrave on the northern eaves adds a delicate style to this grey-toned, cement stucco apartment building.

The stairwell is particularly amazing due to how the architect was able to creatively make use of a rather odd, hexagon-shaped base.

Made of grey cement and yellow terrazzo, the stairwell appears especially captivating from the top floor when looking down.

The five-story building is likely the work by Leonard, Veysseyre & Kruze, a prolific French firm in the city that had a big impact on the look of the former French Concession.

The French architectural firm designed numerous buildings around the city such as the classic Cercle Sportif Francais (the annex of the Okura Garden Hotel) and the Bearn Apartments (Shanghai Women Goods Store) on Huaihai Road.

The firm's influence on and contribution to modern Shanghai architectural design is impossible to ignore. On July 14, 1934, the French newspaper, *Le Journal de Shanghai*, published a supplement showing more than 60 buildings designed by the firm along with the firm's group photo and a map of the former French Concession.

They achieved great success by promoting a new style through their many and varied architectural achievements. Largely owing to their work, the former French Concession had kept up with global architectural trends. As one of Shanghai's earliest practitioners of Art Deco, the French firm designed a series of Art Deco apartments, including the Willow Court Apartments on nearby Fuxing Road.

The Midget Apartments also reflect the Art Deco trend that swept across the city in the late 1920's and into the '30s. However after studying original drawings of the Apartments, Tongji University professor Zheng Shiling finds it was a work of the design team from Credit Foncier d'Extreme-Orient that is largely responsible for the structure's design.

The China Press report proved his study. "It stands on the smallest area of land apartment house in the city. Built by the Credit Foncier d'Extreme-Orient and a Chinese owner, the house is four stories high with flats of four and two rooms. Indirect lighting effects and modernistic interior decorations add to the exotic appearance of this building," the 1931 picture story notes.

It's interesting that Arthur Kruze, one of the French firm Leonard, Veysseyre & Kruze's three partners, was a resident, living in a flat on the fourth floor.

According to the 1930s and 1940s *Shanghai Directory*, most residents in the Midget Apartments were expatriates working for large Western companies.

Residents included R. F. Pirard, who worked for Sterns Ld. China Agency, which sold lubricating oils and greases. V. B. Russakoff, another resident, was employed by petroleum company Texaco. R. Berg was a staff member of engineering company Telge & Schroeter.

All three companies were located on or near the Bund. It was common at the time for Shanghai expatriates to work in that area while living in the quieter western district of the former French Concession.

The three partners of the French architectural firm experienced different lives afterwards. Kruze, who had lived on top of the Midget Apartments, left Shanghai and was never heard of. Paul Veysseyre also left Shanghai and designed many buildings in Vietnam. Based on his well-preserved archives, Veysseyre's biography was published in recent years.

However the firm's most talented Alexandre Leonard went missing in Shanghai during the turbulent 1940s. News stories on old Shanghai newspapers traced ups and downs of his life in Shanghai, where he enjoyed a successful career. After the death of his wife from illness, he remarried and then seems to have vanished into thin air after 1941. His grandson in France is still looking for the whereabouts of this legendary architect.

Tips

Please admire the Art Deco details on the façade such as the geological relief, architectural details shaped like stairs, and cloud-shaped architraves.

HN-262
庭院深深的湖南别墅
Villa with a Deep Garden

漫步武康路湖南路口会看到白墙间的一扇黑色大门，总紧闭着，颇为神秘。

这就是著名的湖南别墅，从不对外开放。只能从史料里了解几任主人的传奇故事。

建筑是典型的西式花园住宅，灰墙尖顶，"人"字形山墙，偌大的花园，与附近的巴金故居有几分相似。住宅底层是会客空间、餐厅、厨房和面向花园的敞廊，卧室和露台在二层，顶层还有阁楼。

住宅建于1931年，历史建筑铭牌提到这里原是英商锦隆洋行大股东的住宅。锦隆洋行是早在19世纪就进入中国市场的英国保险公司，原名"Attorneys of The Excess Insurance Company Limited"。洋行和知名建筑事务所公和洋行一样，到中国后也起了"锦隆"这个好口彩的名字。

1943年，投靠日伪政府的政客周佛海搬进别墅，研究法租界的同济大学副教授刘刚认为，从此这个花园住宅成了一个政治空间。

1949年5月27日上海解放，6月4日英文报纸《密勒氏评论》就刊登整版关于"上海新市长"陈毅的人物特写，介绍他留学法国和后来参加革命的经历，还配了一张陈毅英姿勃发的肖像照片。

同年夏天，陈毅和邓小平两家都搬入湖南别墅。两家合住只有短短数月，但在楼前留下一张珍贵的合影。两人坐在藤椅上，夫人子女环绕身旁，十分温馨。

据陈毅长子陈昊苏回忆，邓家1949年9月赴京后，陈家继续住在别墅，后来又搬至兴国宾馆。在他的童年记忆里，"湖南别墅的院子很大，有很多花，我们在花下照相。"

1962年，陈毅调离上海，他安排毛泽东第二任妻子贺子珍从泰安路搬到安静的湖南别墅。根据贺子珍侄女贺小平的回忆，贺子珍入住湖南别墅后基本处于独居状态。由于进入人员限制，加上贺子珍作息时间与常人不同，她和老战友联络甚少。在庭院深深的湖南别墅里，这位"前第一夫人"度过了一段沉寂的岁月。

参观指南

建筑不对外开放。

The big black gates of Hunan Villa are always closed. The history of the building can be told only through archives.

The villa mirrors the adjacent Ba Jin Residence in many ways. It's a Western-style garden villa with grey walls, a sloping roof and a large, deep garden. The ground floor features a sitting room, dining room, kitchen and a loggia facing the garden while bedrooms, a terrace and loft are on the upper floors.

The name plate of the building says the 1931 house was formerly the residence of a major shareholder from "Attorneys of The Excess Insurance company Limited," a British insurance company which entered the Chinese market in the 19th century.

In 1943 Zhou Fohai, a high-level Chinese official who surrendered himself to work for the Japanese puppet government moved into the villa and had the residence re-purposed as a political space.

On June 4th, 1949, *The China Weekly Review* published a full-page profile on Shanghai's new mayor Chen Yi. Chen's family and Deng Xiaoping's family moved into Hunan villa in the summer of the year. The two families shared the home for only a few months but left a precious group photo of themselves in front of the building. Chen and Deng sat on wicker chairs, surrounded by their wives and children.

According to Chen Haosu, Chen Yi's elderly son, the Chens continued to live in the villa after the Dengs moved to Beijing in September, 1949 and later moved to the Xingguo Hotel nearby. "Hunan Villa has a large garden with many flowers. We took pictures under the flowers," he recalls.

Chen Yi left Shanghai in 1962 and arranged He Zizhen, the second wife of Chairman Mao Zedong to move from her Tai'an Road home to the quiet Hunan Villa.

Her niece He Xiaoping recalls that her aunt lived in almost solitude here and had little contact with old friends. It was at this deep garden at Hunan Villa that the former Madame Mao spent the last lonely years of her life.

Tips

The building is not open to the public.

WK-129
空留回忆的西班牙小屋
A Spanish House of Memories

武康路129号是一座简洁活泼的西班牙式小屋,《字林报行名簿》记载1935年后小屋的主人是意大利商人德利那齐。

小屋呈现武康路一带西班牙式住宅的特征,米黄色墙面点缀着半圆形券门,屋顶覆盖红色筒瓦。精致迷你的"朱丽叶阳台"与一旁美杜公寓的方形大烟囱相映成趣。

德氏经营一家意大利船运公司,1932年当选上海意大利商会主席。作为一位社交活跃的商界领袖,他参加的活动不时见诸老上海的报章。

1936年12月,意大利大使馆经济商务参赞处在外滩26号扬子保险大楼成立。德利那齐客串意大利驻华大使的翻译,一张他拿着讲稿认真翻译的照片刊登在《北华捷报》上。德氏在武康路的邻居——意大利驻沪总领事内龙也参加了这个活动。

1937年2月4日,时任上海市市长吴铁城在海格路(今华山路)

家中设宴,款待意大利大使,宾客中又有内龙和德利那齐。他们从各自武康路的家步行到市长家赴宴,不过10分钟之遥。

《上海邬达克建筑地图》将129号收录为建筑师邬达克的作品。不知邬达克是否与德利那齐相熟,但他对意大利文化情有独钟。

邬达克年轻时曾到意大利深度旅行,做了大量建筑细节的素描。他独立开业后的力作——宏恩医院(现华东医院1号楼)便具有意大利的氛围和风格。他在中国学习了意大利语,为意大利官方俱乐部设计建筑。1939年,他还应邀赴热那亚觐见教皇庇护十一世。不幸的是,教皇在约定时间的前两天去世了。同年,他又在上海出版的意大利文化杂志《马可·波罗I》上撰写了一篇英文短文《中国拱桥》。

旅居上海武康路的美国学者卡洛琳·罗伯森认为,129号是一座充满回忆的小屋,因为她找到了一段珍贵的回忆。

二战期间,德氏公司的员工扎哈罗夫和母亲被日军赶出淮海路培恩公寓(今妇女用品商店),母女二人被德利那齐收留安置在129号家中。

在她的记忆中,身材高大的德利那齐有一双蓝灰色的眼睛和灰色卷发。"他戴着一副眼镜,看上去很尊贵,也很绅士,总是穿得好看。"她提到129号别墅里有很多美丽的中国古董家具,如宁波床和中国陶塑马,椅子和沙发都饰有漂亮的锦缎。

1943年后,战争夺走了德利那齐的财富。他的船只被日军没收,仓库等不动产在政治动荡的年代大幅贬值,最后离开上海时仅带走一些金条和现金。

而战争同样改变了129号的设计师邬达克的命运。这位曾是一战战俘的建筑师,1918年冒险逃到上海,开创了一番传奇的建筑事业。1947年,邬达克遭软禁后选择再次逃亡,他和家人登上波尔克总统号前往欧洲,同年还到意大利参观了圣彼得大教堂下的考古挖掘工作。

如今,129号静静矗立在游人熙攘的武康路,空留昔日主人和建筑师的人生故事与传说。

参观指南

观察"朱丽叶阳台"与美杜公寓大烟囱既对比又和谐的效果。

129 Wukang Road is a Spanish house in a simple, vivid style. Italian merchant Dr. D Tirinnanzi lived here from 1935 well into the 1940s.

The lovely building facade includes features common to Spanish villas, such as red barrel tiles and white-yellow walls punctuated with an arched gate. A delicate "Juliet's balcony" echoes with a square-shaped chimney of the adjacent apartments building.

Tirinnanzi managed a shipping agent company named Comp. Ital. d'Estrem. He was an active business leader and elected chairman of the Italian Chamber of Commerce for the Far East in 1932.

When the offices for the Commercial Counsellor to the Italian Embassy in China opened in the Yangtsze Insurance Building on the bund in 1936, he not only attended the ceremony, but also translated the ambassador's speech into English. His neighbor on Route Ferguson, Italian Consul General L. Neyrone was also present.

Both Tirinnanzi and Neyrone also joined the dinner hosted by Shanghai mayor Wu Tiecheng for the Italian ambassador Vincenzo Lojacono in their beautiful Chinese-style residence on Avenue Haig (today's Huashan Road) on February 3, 1937. The mayor's house was a brief 10 minutes' walk from their homes on Route Ferguson.

The book *Shanghai Hudec Architecture* listed No. 129 as a work by L. E. Hudec, who became a "Shanghai symbol" in recent years.

It's unknown if Hudec and Tirinnanzi were acquainted with each other, but the Slovakia-Hungarian architect had a fondness for Italian culture.

At his young age, Hudec travelled to Italy for several months where he made hundreds of sketches on architectural details. In 1925 he was successful in adapting an Italian style to the Country Hospital (now Building 1, Huadong Hospital), a masterpiece completed after he opened his own firm. He learned Italian in China and cultivated close friendships with Italian politicians and merchants

In January 1939 he set off for a trip to see Pope Pius XI, who unfortunately died on 10 February, two days before the appointed time. The same year, he published a paper on the theme of Chinese arched bridges in the Italian culture magazine *I Marco Polo*.

The American researcher Carolyn Robertson, another resident of Wukang Road, called No. 129 "a villa of memoirs".

According to Robertson's research, Floria Paci Zaharoff was given the offer to move into the wing to her boss Tirinnanzi's villa after being evicted from the Bearn Apartments in 1944 by the Japanese.

In the former employee's eyes, Dr. Tirinnanzi was tall with bluish grey eyes and matching curly hair. The beautifully dressed gentleman wore a monocle and looked distinguished. His villa was full of antique Chinese furniture including a Ningpo bed and antique Chinese clay horses.

Tirinnanzi lost much of his fortune during World War II as his merchant ships were occupied by the Japanese army and his properties devalued during those turbulent years.

The war also greatly changed the fate of architect Hudec. A former POW from World War I, he escaped to Shanghai in 1918 and established a legendary career as a successful architect and left over 100 buildings including the Park Hotel, once the tallest building in the Far East. However in 1947 he chose to leave for Europe for good where he later enjoyed the extraordinary opportunity to visit the archaeological excavations underneath St Peter's Basilica.

No. 129 is now empty and dusty, keeping the past legends silently on the sometimes bustling Wukang Road.

Tips

Please admire the contrasting yet harmonious effect between the "Juliet's balcony" and the big chimney of the adjacent apartment building.

WK-240

大洋行设计的"小熨斗"
A Mini "Flat-Iron"

开普敦公寓的形状酷似迷你版的武康大楼。如果说武康大楼是个三角形的"大熨斗"，开普敦公寓则宛若一把精巧的"小熨斗"，精确地嵌入所在的地块。

公寓由大名鼎鼎的公和洋行设计自用。这家建筑事务所1868年在香港创立，原来中文名叫"巴马丹拿"。1912年，公司委派年轻建筑师威尔逊到上海开设分所，实力不俗的他运气也好。第一个项目联保大楼（即外滩三号）就一炮打响，公司由此打开局面，连续赢得一系列重要项目，成为当时上海滩最叱咤风云的建筑事务所。

研究武康路的同济大学钱宗灏教授很欣赏开普敦公寓的设计，他发现建筑师最大限度地利用了这块形状特殊的基地。"旧法租界的很多基地都非常小，形状也不规则，但这反而促使建筑师们发挥了能动性，最后展现出极其巧妙的设计构思。像开普敦公寓这样的小型公寓，外观简洁清爽，结构合理舒适，受到都市白领的追捧"。

开普敦公寓建于20世纪40年代，占地面积仅有126.7平方米，是一幢四层现代公寓。建筑立面呈三段划分，米色外墙仅以白色线条装饰，明快流畅。公寓那颇有特色的锐角转角的弧形处理相当巧妙，与沿街立面走势连贯。此外，窗户的造型也活泼多样，圆形、正方形和矩形不等，是点睛之笔。

联保大楼在外滩完美落成后，公司在楼内设立办事处，并根据上海文化习惯取了"公和洋行"这个好名字。年轻有为的威尔逊当上了合伙人，带领洋行在"远东巴黎"大展宏图，竟陆续完成了外滩天际线上最壮观的原汇丰银行大楼、海关大楼和原沙逊大厦等建筑。

走进开普敦公寓，楼梯上铺着红黑马赛克及黄油色瓷砖，不仅保存完好，今天看来依旧时尚。公寓底楼开着一行商铺，楼上仍为住宅。

老上海英文报纸上刊登过公和洋行在淮海路设计的小型住宅，真是小而美，与他们在外滩打造的"高大上"形成鲜明对比。

参观指南

建筑内部不对外开放。逛武康路时很容易错过这座小巧的公寓，建议放慢脚步仔细欣赏它的造型、简洁美和建筑细节，并与武康大楼和密丹公寓作比较，会很有趣。

The Cape Town Apartments on Wukang Road are like a miniature version of the nearby Normandie Apartments.

Known by locals as the "small iron" building because of its shape, the Cape Town Apartments were designed by the firm Palmer & Turner, a Hong Kong firm founded in 1868. The architectural firm entered the Shanghai market in 1912, making a big impression with the Union Building (now No. 3 on the Bund).

The company went on to play a leading role in the reconstruction of the Bund in the 1920s and designed nine of the 23 waterfront buildings including the HSBC Building and Sassoon House.

It was the largest and most important architecture firm in Shanghai in the 1920s and 1930s and their works were almost a mini collection of the Shanghai architectural scene.

Qian Zonghao, a Tongji University professor, finds the building made excellent use of an odd-shaped plot of land, which was common in the former French Concession.

"These types of apartment buildings were constructed in later periods, the 1930s or 40s. They were often built on small, irregular shaped pieces of land close to the street or around the corners, just like pieces of dried bean curd," the professor says.

He explains that the French Municipal Council had forbidden new "Chinese-style houses" from being built in the upscale residential area. Due to land plot limitations, architects abandoned classic or Baroque styles in favor of the Art Deco or modern styles.

"On the other hand, these awkward plots forced architects to be more creative and come up with smart designs. It led to a bunch of clean-cut, chic apartment buildings with scientific layouts just like the Cape Town Apartments. These buildings were popular among the city's white-collar employees," Qian says.

Built in the 1940s, the four-story Cape Town building covers an area of 126.7 m^2 and has a rather flat top. The façade is divided into three sections vertically with the middle part protruding toward the street. The creamy-hued façade is decorated simply but for several white lines. The acute angle of the building is treated in a curved way, which flows smoothly to the roadside façade. The windows are in a variety of shapes including circles, squares and rectangles.

Inside, the original staircase is paved with a stylish red-and-black mosaic and the butter-hued tiles are well-preserved. Most of the ground floor has been re-purposed for commercial use while the floors above remain apartments.

Professor Qian adds, the impression that a big firm like Palmer & Turner would only take on big projects is not always right as the Cape Town Apartments demonstrate. Old Shanghai English newspapers also introduced small villas along today's Huaihai Road designed by Palmer & Turner, which contrasted with their much larger projects on the bund.

Tips

The interior is not open to the public. Have a close look at the acute angle of the building and compare the smart design of this building with the Midget Apartments at 115 Wukang Road, or the Normandie Apartments.

WK-117-1
中西合璧的银行家别墅
A Financier's Villa

　　武康路117弄1号是一座精美的花园住宅，金融家周作民曾在此居住。

　　别墅深锁在一扇镂空大铁门后面，不过高大气派，透过黄色围墙仍然可以欣赏顶部的两层。1988年版《上海近代建筑史稿》（简称《史稿》）刊登过一张全貌照片。

　　别墅建于20世纪40年代，由中国建筑师范能力设计。《史稿》一书认为这是建筑师为迎合居住者的特殊喜好设计的住宅，没有自己的风格，是其他住宅建筑形式的组合。"福开森路（今武康路）117弄1号，外形是一座西式花园住宅，室内却用中国传统建筑的彩画平顶和广漆地板。"

　　书中还举了其他两个类似的例子，即形似中国庙宇、室内却采用西洋装饰的华山路吴铁城住宅，以及大名鼎鼎的马勒别墅。那座挪威住宅风格为主的西方建筑，虽有陡峭的屋顶和尖塔，围墙却用中式琉璃瓦压顶，庭院内有石狮装饰。

周作民故居也是一座标准的花园住宅。正屋朝南，设计有敞廊和阳台。基地方正，宅前植以雪松龙柏，草坪绿化，高墙将院落围成一个小天地。

小天地的主人周作民出生于1884年2月12日，1906年赴日本留学，1912年起先后在南京临时政府财政部和交通银行任职，1917年5月参与创立金城银行，任董事兼总经理。1923年他联合金城、盐业、中南、大陆四家银行，创立实力雄厚的四行储蓄会。1929年创办太平保险公司。

周作民是近代中国金融史上著名的银行家。1922年7月22日，他严肃认真的照片登上《密勒氏评论》的"Who's Who in China"栏目。

1937年5月，金城银行在江西路200号总部礼堂隆重举行成立20周年庆典，1 000多名各界名流参加，包括宋子文、何应钦、张群等政经要人，金融巨子更是如云。

《大陆报》刊登了周作民的回顾致辞。他指出，金城银行自1917年成立以来取得"显著的发展"，资本金额从初创时的50万元增加到超过1 000万元，在20年间翻了20倍。周作民还强调，银行的发展得到股东和商业领袖们的大力支持，但仍受到20年来中国政治和经济动荡的影响，否则业绩会更好。

抗战期间，出于种种原因，周作民与日伪往来频繁，被称为"灰色银行家"。他用蓝色墨水写就的日记记叙了在武康路家中往来会客与庭中散步的情景，与金城银行档案一起珍藏于上海档案馆。

1951年，周作民受邀担任全国政协委员，1955年病逝于上海，终年71岁。

他联合建立的四行储蓄会在1934年投资兴建了一座大楼，就是今天的国际饭店。他创办的太平保险发展为今日的中国太平保险集团（简称"中国太平"），是一家拥有47万员工和总资产近5 000亿元的金融保险集团。

而他80年前发表庆典讲话的金城银行大楼，由华人建筑师庄俊设计，今天是交通银行大楼。也是西方古典风格的大楼，却在会客室采用传统中国元素装饰，与中西合璧的武康路别墅异曲同工。

参观指南

建筑不对外开放，可以在武康路上欣赏二层小楼的外立面。建议前往位于江西路200号的金城银行，以及位于人民广场的国际饭店。

The exquisite garden villa on 117 Wukang Road was the former residence of financier Zhou Zuomin.

Though concealed behind a large iron gate, the upper two floors of the grand villa can still be appreciated through yellow-hued walls. The 1988 book *Shanghai Modern Architecture* published a photo of this building, which was designed by Chinese architect Fan Nengli in the 1940s.

According to this book, the residence was designed in a mixed style according to the preference of its owner. "Building 1, 117 Route Ferguson (now Wukang Road) takes the form of a western garden villa. However, the interior decoration features traditional Chinese painted ceilings and a Chinese lacquer floor."

The book compares it with other two similar examples: politician Wu Tiecheng's residence on Huashan Road and the renowned Moller Villa on Shaanxi Road N. Shaped like a Chinese temple, Wu's villa features Western-style interior decorations. The Norwegian-style Moller Villa includes a steep roof and sharp-pointed tower. The surrounding walls are covered by Chinese glazed tiles while the garden is graced by stone lions.

The layout of Zhou's Wukang Road home is typical for a garden villa with the main building facing south, an open loggia and a balcony. The building is fronted with cedars, cypresses and a spacious lawn surrounded by high walls.

Born on February 12, 1884, Zhou Zuomin went to study in Japan in 1906, and returned to work for the finance ministry of the Nanjing Government in 1912 but changed to the Bank of Communications in 1915. He also participated in founding the then famous Kincheng Bank and served as the general manager in May, 1917.

Five year later, he united four big northern Chinese banks — Kincheng, Yien Yieh, Continental and the China and South Seas banks to found the Joint Savings Society. The co-operative venture grew to become a very influential Chinese financial institution. Zhou also founded Tai Ping Insurance in 1929.

As a famous banker in China's modern financial history, his earnest face was published on the column "Who's who in China" of *The China Weekly Review* in 1922.

In May 1937, Kincheng bank hosted a celebration of the 20th anniversary of its founding in the auditorium of the bank building at 200 Jiangxi Road. More than 1,000 prominent people "representing all walks of life, banking and financial leaders predominating were present", including T. V. Song,

Chairman of the Board of Directors of the Bank of China, and Generals He Yingqin and Zhang Qun.

The China Press published Zhou's report, and note that "the dramatic progress" the bank had made was clearly shown in the fact that the capital had grown form the initial $500,000 in 1917 to more than $10,000,000.

He stated that the growth of the bank was even more impressive given that China had been confronting political turmoil and economic uncertainty during the past two decades. The development of the bank's business was mainly to the credit of its supportive shareholders and other financial and business leaders.

Zhou has also been nicknamed as "the grey banker" by Shanghai historians due to what is perceived as a close relationship with the Japanese puppet government during the Chinese War of Resistance against Japanese Aggression. His diaries, written with a blue-ink pen, record visiting guests and garden walks in the Wukang Road home, and have been preserved in the Shanghai Archives Bureau along with archives of the Kincheng Bank.

In 1951 he was invited to serve as a council member of the CPPCC National Committee and died of illness in Shanghai at the age of 71.

The Joint Savings Society he had

周作民故居　Former residence of Zhou Zuomin

formed built a skyscraper in 1934, which is today's Park Hotel on People's Square. The insurance company he founded grew to be today's China Taiping Insurance Group Ltd., a financial giant with 470,000 employees and a 500 billion yuan in capital.

The Kincheng Bank Building, at which he gave a speech some 80 years ago, is now the Shanghai branch of the Bank of Communications. Designed by Chinese architect Zhuang Jun, the western classic building was graced by traditional Chinese elements in the parlor, which mirrors Zhou's east-meet-west villa on Wukang Road.

Tips

While the villa is not open to the public, after admiring the upper two floors from Wukang Road, I'd suggest visiting Kincheng Bank on 200 Jiangxi Road along with the Park Hotel on People's Square which Zhou had contributed to Shanghai architectural history.

WK-378

武康路上的"城中村"
A Chinese "Urban Village"

小而美的武康庭,阳光树影咖啡香,周末生意极好。

武康庭由五座建造于不同年代的历史建筑组成,包括20世纪20年代的红砖别墅、70年代的白色工厂办公楼,以及到90年代才修建的暖黄色房地宾馆。

设计武康庭项目的是澳大利亚建筑师朱萃华,她的理念是打造一个"中国的城中村",让都市人来度过休闲时光,喝杯拿铁,逛精品小店,再到画廊看看现代艺术。

根据武康路的建筑风貌,建筑师选择了装饰艺术风格的元素来设计武康庭,与徐汇区政府对武康路进行风貌改造的思路不谋而合。这种1925年源自法国巴黎的设计风格,由几何、曲线、放射状等图案组成,被认为是现代主义早期的一种形式。武康庭的窗户、栏杆和一些装饰线条都是这种风格的。

建筑师还提到,她特别重视设

计公共空间，如户外平台和屋顶露台。还保留了五层楼高的大树。希望这里成为如纽约格林威治村那样惬意的社区。无论居民还是游客，都可以享受这份没有被水泥大楼破坏的静谧。

历史图纸显示，武康庭中历史最悠久的红砖别墅设计于1926年，业主是一位女士，名叫"T. C. Quo"，很可能是民国外交官郭泰祺的夫人。其中的白色建筑原是上海仪表局的办公楼。而房地宾馆所在的基地，原是一座美丽的花园别墅，很少有人知道20世纪30年代，这里曾是贝聿铭父亲——民国银行家贝祖诒的家。

这座两层的西式别墅曾有一个树木葱茏的大花园，银行家优越富足的生活可见一斑。可是少年贝聿铭并未享受多少。他在《贝聿铭谈贝聿铭》中提到，身为长子的他与生母感情极深，但1930年母亲却患病去世，当时他只有13岁。此后，"父亲为我和弟妹们找了一户很大的公寓居住。从那时起，父亲就与我们分开生活。我就在那种没有母亲的环境中生活了三四年。"

银行安排因中年丧妻而忧郁的贝祖诒出国散心。他在欧洲结识了年轻的外交官女儿蒋士云，不久两人就结婚了。根据《字林报行名簿》记载，贝祖诒夫妇于1933年搬入福开森路378号的新家。

在黄金的20世纪30年代，贝聿铭年轻的继母经常在378号举办社交活动。1935年4月，《大陆报》刊登了30位中外女士在别墅前的合影，并配文字说明是中国妇女俱乐部在贝夫人"美丽的家"举办午餐会，招待访华的美国经济代表团里的五名妇女代表。

花园午餐会后不久，17岁的贝聿铭便独自启程赴美留学。这位贝家长子没有听从银行家父亲的劝告去英国读金融，而是追随自己内心的选择去美国读建筑。少年贝聿铭失去母亲又与父亲分开生活后，常去大光明电影院看电影、打撞球。在那里，他深受美国电影的影响，也被国际饭店的高度深深吸引，从此开始了做一名建筑师的梦想。

参观指南

这里可以是徜徉武康路开始或结束的地方。坐下来，喝杯咖啡，感受"城中村"和武康路的氛围，或者，理理思绪静静心。

When Chinese Women Entertain

1935年4月，中国妇女俱乐部在贝夫人"美丽的家"举办午餐会
The Chinease Women's Club, 1935

The Ferguson Lane project on Wukang Road has long focused on creating a Chinese urban village where people want to spend their leisure time sipping lattes, shopping in trendy boutiques and browsing modern art in the many nearby galleries.

The area comprises five buildings including a 1970s industrial building, the Fangdi Hotel built in the 1990s, and, as its centerpiece, a 1920's villa.

Alexandra Chu, chief architect of the Ferguson Lane project, says "it was one of the best locations in Shanghai, a lot of history and very residential." She recalls when the project just kicked off, "there weren't many shops, very quiet, with the beautiful scale of the street, the trees and old houses."

The Australian Chinese architect opted for an Art Deco style during the revamp. Emerging at the beginning of modernism, the Art Deco style fit quite well with the buildings in the compound and Wukang Road's historical context. Today many of the windows, railings and some of the moldings on Ferguson Lane are Art Deco inspired designs.

The design team also made an effort to emphasize public spaces and Ferguson Lane includes outdoor decks and terraces. They kept some beautiful old trees which reach as high as five-stories, envisioning the neighborhood eventually being Shanghai's equivalent of Greenwich Village in New York or Minami Aoyama in Tokyo. Now it's destination for both residents and visitors to enjoy dining and entertainment experiences in a tranquil setting unspoiled by glass and concrete skyscrapers.

Archival drawings indicate the villa was designed in 1926 by the Southeastern Architectural and Engineering Company for Mrs. T.C. Quo, who was likely wife of a Kuomintang diplomat. The white industrial building was originally the office of the state-owned Shanghai Meter Factory.

It's rarely known that Ferguson Lane's Fangdi Hotel was built on the site of former residence of Chinese financier Tsuyee Pei, the father of famous architect I. M. Pei.

Back in the 1930s, Pei's residence was a beautiful two-floor western villa facing a big, lush garden, appropriate at the time for an affluent financier's family. Unfortunately the architect Pei did not himself spend much time there.

In Gero Von Boehm' book *Conversations with I.M. Pei: Light is the Key*, the architect recalls his profound feelings for his mother, who unfortunately died in 1930 when he was only 13 years old. Since then, Pei's financier father chose to live separately with his children who rent a big apartment for Pei and his siblings.

During a trip to Europe, Tsuyee Pei met the young daughter of a Chinese diplomat and they married soon after. According to *Shanghai Directory*, the new couple moved into the love nest on 378 Route Ferguson in 1933.

In the "golden 1930s," I. M. Pei's step-mother often hosted social events at the villa. In the April of 1935, *The China Press* published a group picture of 30 foreign and Chinese women on the front lawn. They were members of Chinese Women's Club who entertained five women members of the American Economic Mission to China "in the beautiful home of Mrs. Tsuyee Pei on Route Ferguson."

Not long after the reception, 17-year-old junior Pei left Shanghai for the U.S. He did not accept his financier father's advice to study banking in the U.K. Instead, this young man chose to follow his own heart and study architecture in the United States.

After losing his mother and living separately with his father, I. M. Pei often went to the Grand Theatre to

watch movies and play billiards. There he fell in love with American movies and with architecture when he watched the workers digging out the foundation of the Park Hotel and constructing the 22-story structure.

Architect Chu says she enjoys coming to the "urban village" on Wukang Road for coffee on weekends and watching people really enjoy the space.

"They don't know why but they like it," she says. "And it has really changed the character and brought more life to Wukang Road. The growth around the area is still very respectful to the history."

Tips

Pick a nice day and have breakfast or coffee there. It's fun to find the Art Deco details sprinkled around this project.

WK-390

酝酿桑塔纳的地中海别墅
A White Villa

武康路390号是一座沪上少见的地中海风格白色别墅,虽不对外开放,但常引人驻足停留,隔着花园眺望欣赏。

红瓦白墙的别墅由比利时义品放款银行建筑部设计承建,东、西、南三面由一条优雅敞廊环抱。敞廊面向青葱美丽的大花园,景致无敌。南立面有小巧的白色阳台,饰有山花。老虎窗微微伸出红瓦屋面。

这座花园洋房建于20世纪20年代,常被称作"意大利总领事官邸",因为意大利驻沪总领事内龙曾在此居住。也有人因此推测,其建筑风格源于这位地中海国家外交官的品位。

不过根据《字林报行名簿》记载,内龙只是1936年在这里短暂居住过。地中海别墅的主人更迭频繁,几乎清一色为外国侨民,其中居住较长的是从事外汇经纪的安德鲁斯先生。

1927年1月29日,《北华捷报》

报道了一则盗窃案新闻。当时安德鲁斯夫妇刚从国外度蜜月归来不久，位于390号的家中即不幸遭遇了"精心策划的盗窃"。

案发当晚，这对侨民夫妇外出就餐后又去参加舞会，回家很晚所以并未察觉异样，直到第二天早上醒来才发现夫人的衣柜已被洗劫一空。安德鲁斯夫人在巴黎新购的所有长袍和法式内衣，以及数不清的珠宝首饰，甚至包括童年时代的礼物和纪念物品，被悉数盗走。

《字林报行名簿》还显示，这对安德鲁斯夫妇不知为何曾在1936年搬走，但1939年又搬回白色别墅，到1941年时屋主已经变为他人。据说，二战后安德鲁斯夫妇定居美国马里兰的郁金香山，他们离上海万里之遥的新家也是一座历史建筑。

小楼如今由上海汽车工业总公司使用。1983年，这座地中海别墅

见证了中国汽车工业历史上的重要事件。上汽集团与德国大众长达5年的艰苦谈判在武康路390号终获成功，上海桑塔纳轿车由此诞生。上汽集团后来成功走出一条利用外资、引进技术的道路，为中国轿车工业发展提供了一个极其成功的模式。

参观指南

建筑不对外开放，可以在武康路门口欣赏建筑立面、敞廊和大花园。

白色别墅　The white villa

The white villa at 90 Wukang Road was designed in a Mediterranean style which was rare in Shanghai.

Covering an area of 420 square meters, the villa stands out with its snow-white walls, dark wooden beams and red-tiled roof.

The brick-and-wood building is shaped like a huge fan and features seven arches in the front. An artful colonnade caresses the eastern, western and southern facades, offering nice views of the lush garden. The southern façade has a small, white balcony with a curved gable and dormer projecting out from the red-tiled roof.

The villa was once the home of Italian Consul General Luigi Neyrone during the 1930s. Although it was widely assumed that the house was built to his wishes, that was not actually the case.

According to the *Shanghai Directory*, Neyrone moved into the house in 1936, years after it was built in the late 1920s. A long-time resident of the villa was Lewis R. Andrews, a partner with the foreign exchange brokerage firm Wentworth, Andrews & Giese.

On January 29, 1927, the *North-China*

Herald reported a "pre-conceived and carefully arranged robbery at the residence of Mr. and Mrs. Lewis Andrews, 390 Route Ferguson where they had recently moved."

The Andrews couple had recently returned to Shanghai after extensive travels in America and Europe as a part of their honeymoon. They returned home after dining out on a Wednesday evening with subsequent dancing, noticed nothing unusual, and discovered the robbery the next morning.

"Paris gowns and trinkets of considerable value, accounting in all to about Tls. 1,500 were stolen....All of Mrs. Andrews' Paris gowns which she had brought back with the expectation that they would last until another visit to Paris at some future date were taken as part of the loot. Every piece of French lingerie in her bedroom and numerous small pieces of jewelry, many of which were childhood gifts and keep-sakes, a watch and several pendants, were also made off with by the thieves," the report says.

The *Shanghai Directory* notes that the Andrews moved away in 1936 before Mr and Mrs Raymond S. Kin moved into the house in 1938. The Andrews moved back into the white villa in 1939, but they didn't stay long. Ownership of the house changed to A. Sadoe in 1941. It is said that after World War II the Andrews moved to a historic house called Tulip Hill in Maryland, USA.

The building is now used by the Shanghai Automotive Industry Corporation to host international meetings. The villa's interior was renovated in 2003 to match its new purpose as a commercial meeting center.

In the 1980s, SAIC used the villa as its office. Negotiations between SAIC and Germany-based Volkswagen were held in the villa, the result of which eventually leading to the production of Santana vehicles in China.

Tips

The villa is not open to the public, but its façade and garden can be seen from Wukang Road.

WK-393

武康路驿站的双重空间
A Historic Stop on Wukang Road

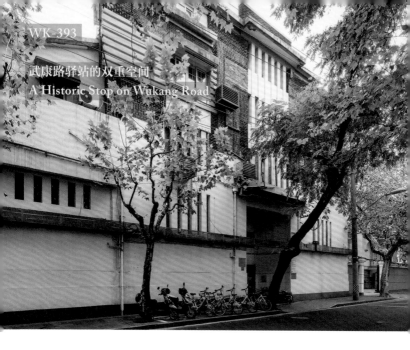

393号在武康路上有两扇大门。

黑色铁门里是近年成为武康路驿站的徐汇区老房子艺术中心,这座装饰艺术风格的米色大楼曾是世界社国际图书馆。如果有幸叩开393号的另一扇绿色铁门,会发现一座古典风格别墅深藏其中,这里是"黄兴故居"的由来。两座共用门牌号码的建筑紧密相连,讲述了近代中国的一段历史。

1916年11月的《北华捷报》刊登了黄兴逝世的新闻。报道提到,这位与孙中山并肩战斗的革命领袖身体有恙,已住院治疗多时。最后他被送回位于福开森路393号的家,度过了生命中最后几天时光。

如今,"黄兴故居"已成为二十几户人家的共同家园,内部拥挤杂乱,但仍然可以从红瓦屋顶、灰色卵石墙、绿色窗棂和棕色壁炉中看出昔日的美丽。

1929年,国民党元老李石曾来

到393号，他和张静江、吴稚晖等人在巴黎创办的"世界社"也落户在此。世界社是个传播启蒙思想的非政府组织，通过出版报纸和杂志推动社会变革。

李石曾是保守的晚清重臣李鸿藻之子，但他的思想却相当新潮。1902年，李石曾成为"中国留法第一人"。他在法国巴斯德学院利用现代化学知识研究大豆之后，在巴黎创办了豆腐工厂，成为影响深远的勤工俭学留法运动的前奏。

1933年，国民政府出资在黄兴故居北侧修建装饰艺术风格的图书馆，后来世界社又在393号创办了法语特色的世界学校。

在黄金的20世纪30年代，393号成为一个繁忙的文化中心，活动频频见报。

1933年2月，世界社在此接待英国著名作家萧伯纳；3月，又宴请国际劳工局副局长莫列德。著名法国女飞行员希尔兹中途加入了这场盛大的晚宴，她是从东京返法途经上海，受到时任上海市长吴铁城的接见。

三个月后，一个大型书展在393号举行。李石曾从各类文化机构收集了约10万册关于中国的图书在此展示，另有大量书籍运往日内瓦国际文化研究图书馆。《大陆报》报道，近200位各界精英参加了开幕式，包括中国银行总裁张佳璈、商业银行总裁徐新六、法租界"名流"杜月笙、京剧名家梅兰芳和著名画家刘海粟等。

时光流转到2009年，上海市旅游局为迎接上海世博会学习国外大城市经验，在各区开设游客咨询中心，武康路393号列选。徐汇区旅游局调研后结合武康路特色，为393号挂上了"徐汇区老房子艺术中心"的牌子，常年举办文化讲座和展览，介绍上海老房子的故事。历史仿佛轮回，393号重新成为一个文化中心，它承载的双重历史还在不断增加和丰富。

经过武康路393号，可以放慢脚步，回想这个门牌号码背后叠加的空间和历史。这里曾经是一位革命领袖的家、一个充满理想主义的文化机构、一间展览馆、一所学校……今天仍是一个充满文化和记忆的地方。

参观指南

这个武康路驿站是老房子爱好者的天堂。这里提供探索武康路的步行地图、书籍、多媒体介绍和自行车租赁等。可以在这里规划一下步行武康路的路线。

393 Wukang Road is composed of two distinct parts built in two different eras for different uses. The older part is a four-story wood-and-brick villa built between 1912 and 1915. The newer part is a four-story reinforced concrete structure in the Art Deco style.

Huang Xing, a respected revolutionary pioneer in China's modern history lived in the original house for a short time in 1916. Huang had worked with Dr. Sun Yat-sen to overthrow the Qing Dynasty (1644–1911). He only lived in the villa for several months and died there of disease at the age of 42.

The newer part was built in 1933 for Shijie She, or The World Research Institute, an influential cultural organization founded by Chinese intellectuals in Paris to promote revolution, science and new culture. The Xuhui District government has converted the newer building into a special tourism information center.

The idea for the center began in 2009 when the municipal tourism bureau decided to copy the practice of cities around the world and establish visitor's centers around Shanghai for the World Expo in 2010.

The addition was meant to offer tourists a place to stop on Wukang Road. After a survey, Xuhui district government focused on the character of Wukang Road — "old house" and gave this place a new name — Xuhui Historical Building Art Center. The restored facade features narrow slit windows, pale yellow plaster and grayish-brown brick.

While the revamp has breathed new life into the building, the complicated history and entirely different architectural styles make it an interesting place to visit.

Today the older section now houses more than 20 families. The villa's once spacious grand corridors remain although much of this public space is now used as kitchens or storerooms. Nonetheless, there are still numerous intricate details and passageways revealing its grand past.

With a roof made of red tiles, the house has gray ovals on the exterior walls and a big garden on the south side. The wooden staircase with ex-

quisite patterns is dusty, but still in good condition.

The original villa was designed in a neo-classical style and included a parlor, dining room and studio on the second floor and four large en suite bedrooms on the third.

After Huang died in 1916, the villa was repurposed over the years as an exhibition hall, a library and a school by Shijie Xuexiao (The World School) which recruited children of intellectuals to prepare them for further education in France.

The main founder of the institute, famed socialist and educator Li Shizeng, had founded a Toufu company in Paris after studying soybeans in the Pasteur Institute.

He was also noted for his work on the famous "Work-Study Movement" that sent many young Chinese to study in France beginning in 1919. The movement provided China with several future leaders including Zhou Enlai and Deng Xiaoping.

In the "golden" 1930s, No. 393 was a busy venue for cultural events and receptions.

In 1933, the society received British author and playwright George Bernard Shaw and famous French female pilot Mile Hiltz.

Later that year the exhibition of the China section of the International Cultural Research Library at Geneva opened here. More than 200 local government and civic leaders visited the literary exhibition including Zhang Jia'ao, general manager of the Bank of China, renowned Beijing opera actor Mei Lanfang and renowned painter Liu Haisu.

At present, the two-floor visitor center regularly hosts exhibitions and activities that are often related to old buildings. Guests can download pictures through Bluetooth, read historical documents or rent bicycles to more fully explore the neighborhood.

The building which was once a revolutionary pioneer's home, an exhibition hall, a library, and a school is now a place for them to rediscover stories of yesteryear.

Tips

The center provides many services that an old house-lover would need to explore the Wukang neighborhood, from maps, brochures, books, bicycles and bi-lingual staffs to answer questions. If you understand some Chinese, you can also join its free-of-charge story-telling events which often include Shanghai-style cultural performances.

WK-395

水晶与玫瑰
Radium and Roses

武康路395号是一座巴洛克风格别墅。红瓦坡屋顶,立面饰有弧形大阳台,布满曲线。丰富多变的建筑细节还包括椭圆形屋顶窗、塔司干柱式和窗户上的拱顶石,巴洛克建筑的奢华精细让人过目难忘,与马路对面地中海风格的390号别墅相映成趣。

历史建筑铭牌显示建造年代为1926年,西文报纸较早提及395号是在1928年。《北华捷报》在当年2月8日报道,已故西人德拉恩菲尔斯的女儿尤金妮亚小姐与沃茨先生在外滩联合教堂喜结连理。仪式结束后,新娘的母亲在福开森路395号家中举行招待会,款待参加婚礼的众多友人。

1933年3月6日,《大陆报》刊登新娘一家将在395号拍卖家具地毯的预告。此后数年,上海行名录里没有关于395号的信息,直到1937年显示这里变成"镭学所"。

镭学所是国立北平研究院的八个研究所之一。国立北平研究院由李石曾提议创办并担任院长，1929年成立于北平中海怀仁堂。其中镭学所和药学所于抗战前迁到上海福开森路395号。

1948年国民政府行政院出版的《北平研究院》一书提到，镭学所内"设有放射学X光、光谱学等研究室及化学实验室，又设有图书室，金工修配场等。"抗战初期由于运输困难，镭学所未能随军西移，留沪暂行工作；之后太平洋战争爆发，租界被占领，研究工作始告停顿。抗战末期395号曾被敌伪攫夺，"幸仪器、图书、药品散失尚微，但因弃置日久而损坏者，则颇不少。"而药物所迁到395号时，仪器有检光计、显微镜和微量分析仪器等，还有足够使用数年的欧洲名厂药品。与抗战期间在京遭受重大损失的其他几所相比，迁到武康路的两所已经幸运多了。

战火纷飞的岁月里，科研人员坚持研究，取得不少成果。镭学所成功仿造德国哈瑠维氏医用紫外光灯，又获得水晶仪器制造的新技术，伦敦大英自然博物馆也发来合作邀请。1937—1947年，药学所在国内外知名期刊发表了27篇研究论文。

1950年后，两所并入中科院，武康路395号先后用作厂房和演员剧团，近年历经两次修缮，现为办公用房。

仔细翻阅20世纪20年代的西文报纸，395号新娘尤金妮亚的上海人生由几则新闻勾勒而出，与研究院的战时岁月同样跌宕起伏。

她1905年出生于俄国西伯利亚，1920年随父母抵沪，1922年年仅44岁的父亲不幸去世。两年后，她与美军上尉亨特结婚。1927年，一身白衣的她到美国法庭起诉离婚，自述婚后与丈夫争吵不断。两人前后赴菲律宾和美国旅居，但丈夫劝她回到上海母亲身边，并不再与她联系。

1928年，这位命运多舛的俄国小姐终于再次披上婚纱。关于她的婚礼报道这样写道："新娘美丽动人，穿一件饰有银珠与纱罗的法式象牙色缎面礼服。她搭配了一双银色鞋子，手上捧着一大束白玫瑰与山谷百合。"

参观指南

建筑不对外开放，建议欣赏立面上巴洛克风格的细节，如布满曲线的立面、椭圆形屋顶窗和窗户上的拱顶石等。

A villa in the Baroque style, 395 Wukang Road features a red-tiled sloping roof, a façade adorned by curved balconies, Tuscan orders and windows decorated with keystones. This Baroque-style building and the Mediterranean-style No. 390 are two showpieces on the southern part of Wukang Road.

The name plate of No. 395 says the villa was built in 1926, which was reported by a local English newspaper in 1928 with a piece of wedding news. Mrs. Eugenia Hunt, daughter of Baroness and the late Baron W. W. Drachenfels became the bride of Lieutenant Lawrence David Watts of the Machine Gun Co., S.V.C. at the Union Church. After the ceremony, a reception was held at the residence of Mrs W. W. Drachenfels at 395 Route Ferguson, "where many friends met to offer their congratulations."

On March 6 1933, the building was mentioned again in a notice in *The China Press* regarding an auction of furniture and carpets. It's unknown where the Drachenfels had moved to, but No. 395 was absent from the *China Hong List* in the following years.

In 1937 the residence became the "Institute du Radium", one of the eight institutes of the National Academy of Peiping, which was founded by famed socialist and educator Li Shizeng in Beijing in 1929. Two of the academy's institutes, radium and medicine, were moved to 395 Route Ferguson before the War of Resistance against Japanese Aggression. And Li also founded the World Institute in the adjacent No. 393 building.

According to the 1948 book *National Academy of Peking*, the Institute of Radium was well-equipped with "laboratories of radiology, spectroscopy and chemistry as well as a library and a medal workshop. Each lab is provided with water, electric and gas supply." The facilities included a platinum pipe containing 57 milligrams of radium, a Curie electrometer, chemical medicines and some 1,000 books.

After the war broke out, the Institute of Radium had to maintain its operation in Shanghai since it was unable to relocate to Western China along with Chinese military due to the inherent difficulty in transporting such technologies and materials.

Research at the lab did not stop until the former French concession was occupied by the Japanese army after the Pacific War began in 1941. Despite the fact that No. 395 had been taken over by the Japanese puppet government, most instruments, books and medicines remained well preserved. However, some facilities were damaged, having not received proper maintenance for several years.

The Institute of Medicine brought appliances like photometers and microscopes to No. 395, as well as an abundant amount of medicines produced by famous European factories. The two institutes were far more fortunate than the academy's other institutes which were badly damaged in Beijing during the war.

During the war, Chinese scholars continued their research in this Baroque villa. The Institute of Radium successfully produced medical ultraviolet light among other new appliances for manufacturing crystal instruments. Even the Natural History Museum in London sent an invitation for cooperation with the Chinese institute. The Institute of Medicine published 27 papers on domestic and overseas journals.

The two institutes were later incorporated into the Chinese Academy of Science. After that No. 395 was repurposed as a factory and, later, a theatrical troupe It is currently used as an office building.

While in Shanghai, the life of the bride Eugenia herself was also filled with many turns and twists, not unlike the history of the institute during the period of military occupation.

Born in Russian Siberia in 1905, she arrived in China with her parents in 1920 and lost her 44-year-old father in Shanghai in 1922. Two years later, she married Lieutenant Ralph B. Hunt, U.S.S at the Russian Orthodox Church. In 1927 Eugenia, "a slim, tall figure dressed in white" went to the American Court to file for divorce. She said

李石曾　Li Shizeng

they had always been arguing and that they had grown increasingly apart. He sent her home with her mother and stopped replying to her letters.

One year later, this Russian girl was once again attired in a wedding gown.

The 1928 news has described her second wedding in great detail.

"The bride looked very charming in a robe de style of heavy ivory French satin trimmed with silver beads and tulle maline. Silver shoes and stockings were worn and a big shower bouquet of white roses and lilies of the valley completed the costume."

Tips

The building is not open to the public. Please admire its Baroque features, including the curved façade, oval-shaped roof windows and keystones above the windows.

HH.M-1843

宋庆龄的船形别墅
Madam Song's Ship-Shaped Villa

宋庆龄故居的白色别墅形似一条大船，有说法原主人是"希腊船主鲍尔"。其实这位鲍尔是个美国引航员。

砖木结构的别墅占地700平方米，朝南面向长方形的大花园，草坪四周绕着绿色香樟。两层共有八个房间，包括起居室、餐厅、书房和卧室等，基本保持了宋庆龄居住时的原貌。细心的参观者会发现绿色木窗上有小巧的船锚图案，烟囱好似桅杆，这可能与原主人鲍尔的职业有关。

20世纪20年代，美国记者库恩曾在上海工作，她撰文回忆上海总会长吧时提到过老上海引航员这个特殊的职业。

这条超过100英尺长，由抛光的桃心木制成的长吧据说是世界上最长的吧台。透过酒吧的凸窗可以看到黄浦江忙碌的港口。长吧的好座位都留给扬子江的引航员们，因为是他们驾驶船只巧妙地绕过暗流和浅滩,带领人们从入海口来到上海。

1929年10月20日,库恩供职的《大陆报》刊登了著名引航员鲍尔病逝的新闻。这位引航员当时年仅46岁,因腹膜炎在邬达克设计的宏恩医院(今华东医院)去世。鲍尔1883年出生于美国罗德岛,有丰富的航海经验;1918年加盟上海注册引航员协会,11年来一直勤奋工作,将远洋游轮导引到上海港。他也是上海美国总会和哥伦比亚俱乐部会员,同时还是共济会32级会员。

1928年《密勒氏评论》的一则新闻透露了引航员是收入丰厚的垄断行业。鲍尔所属的协会只雇佣外籍引航员,一开始只招收英国人,一战后也开始吸纳美国人等其他外国侨民,但没有中国引航员。

根据宋庆龄故居纪念馆档案,别墅后来多次转手,二战后国民政府将小楼作为敌伪产业处理,用于接待外宾。

宋庆龄战后回到上海,发现香山路旧居水管被日本人破坏,于是被政府安排到桃江路小楼居住,那里潮湿闷热,加重了她的皮肤病。1949年,宋庆龄搬入这座闹中取静

的别墅。她还继续使用原来室内的家具,只带入一套嫁妆家具。

故居纪念馆宣教中心的傅阳主任提到,1950年,为了安全考虑,要给别墅围墙安装电网。但是宋庆龄不同意,她说附近的小猫小狗进来也是客人,担心电网会伤害它们,所以后来改成了竹篱笆。

白色船形别墅成为宋庆龄一生中居住时间最长的房子。虽然1949年以后她经常去北京开会,在京也有住所,但她对上海有特别的感情,个人书信里只称上海的家是"我家里"。

参观指南

故居开放时间为上午9点到下午4点,可以寻找一下原主人引航员留下的航海元素。

Song Chingling's white villa on Huaihai Road is shaped like a gigantic boat, which was probably owing to its original owner, ship pilot Leo. R. Ball.

Covering an area of 4,330 square meters, the 700-square-meter villa is a brick, wood and concrete structure facing a large lawn surrounded by tall camphor trees.

The two-story European-style villa has eight rooms including a sitting room, dining hall, study and several bedrooms. The chimney for the fireplace resembles a ship's mast. The green wooden blinds are carved with tiny, delicate patterns of Chinese junks and anchors, likely a reference to captain Ball's profession.

Irene Corbally Kuhn, a foreign correspondent in Shanghai in the 1920s, recounted pilots and the renowned long bar in the Shanghai Club in an article "Shanghai: The Way It Was."

"More than 100 feet of dark, polished mahogany, it was said to be the longest bar in the world. A wide bay window in the barroom overlooked the frenzied harbor traffic. Tables there were commonly reserved for that colorful breed, the Yangtze River pilots, the men who negotiated the tricky passage through shoals and sand bars from the estuary to Shanghai and beyond," she wrote.

On October 20, 1929, *The China Press* which Kuhn worked for published the news of Ball's death at the Country Hospital (today's Huadong Hospital) after losing a long battle with peritonitis.

"The late Capt. Ball was born in Providence, Rhode Island, U.S.A. in 1883 and, after gaining his early experience on the seven seas took an active part in the building of the Panama canal under General Gorthals…In 1918 he resigned the command of the S.S. China to join the Shanghai licensed pilots' association and for the past eleven years has been continuously engaged in piloting other trans-Pacific liners into the port…He was a member of the American club, Columbia country club and French club; he also was a 32nd degree mason," the news reported.

Another story in *The China Weekly Review* on October 6, 1928 reported that members of The Shanghai Licensed Pilots Association were composed exclusively of British pilots with just a few Americans. The profession seemed to attract seasoned, adventure-seeking individuals looking for well-paid work after World War One.

According to archives of the Song Ching-ling Memorial Residence, Ball's family sold the house in 1930, which had by then been owned by several people including renowned Shanghai financier Zhu Boquan. The Kuomintang government confiscated

it after 1945 and used it temporarily as an official guesthouse to receive American guests.

Song Chingling and her husband, Dr. Sun Yat-sen's previous villa on Xiangshan Road, has since become a museum for Sun. When Song returned to Shanghai from Chongqing after World War II, she found her house had been partially damaged by the Japanese. She was then assigned to live in a smaller house on Taojiang Road by the Kuomintang government but complained of the humidity and the heat in the poorly-built house.

She then moved into this villa at 1843 Huaihai Road M in the spring of 1949. Owing to a tight budget she continued to use furniture left by the previous owners except for a set of bedroom furniture she had received as part of her dowry.

In 1950, the government renovated the villa for security reasons, but Song refused to install an electric fence on its surrounding walls.

"She said the cats and dogs entering the garden were her guests and she feared they might be hurt by an electric fence. A bamboo fence was used instead," says Fu Qiang, director of education department of the memorial residence.

Song lived here until she moved to Beijing in 1963, when she was ap-

宋庆龄　Song Chingling

pointed honorary vice chairperson of the People's Republic of China. From then on she spent time in both cities each year until her death in 1981, when the residence became a museum commemorating her life.

Of the many places she lived, the white, ship-shaped villa was Song's home for the longest period in her life. Though she had another home in Beijing, she always loved Shanghai and in private letters Song only referred to this Shanghai villa near Wukang Road as "my home".

Tips

Opens from 9am to 4pm. Guests especially enjoy exploring the nautical-themed motifs around the villa.

145

HH.M-1850

一枚上海符号
A "Shanghai Symbol"

武康路上最有名的建筑莫过于武康大楼。上海市旅游局近年评选的"上海99个经典符号"名单中,邬达克设计的武康大楼与国际饭店入选,被列为"喜爱上海的理由"。

武康大楼是法商万国储蓄会投资兴建的高级公寓,原名"诺曼底公寓",是邬达克在美商克利洋行时期的重要作品。

万国储蓄会1912年由西人盘腾等人创办,总部设在上海。利用国人喜爱"有奖储蓄"的心理,储蓄会生意兴隆,1934年鼎盛时曾吸纳全国五分之一以上的储蓄额。

储蓄会除了将资金用于投资债券和外汇外,还从事获利丰厚的房地产业,旗下地产公司中国建业地产公司开发了旧法租界的地标建筑,如毕卡第公寓(衡山宾馆)和培文公寓(上海市妇女用品商店)等。

武康大楼巍然耸立的视觉效果源于它的位置,三角形的基地位于五条马路交汇的岔道口，视野开阔，气势好似一艘战舰。这座八层高的大楼是沪上最早的外廊式公寓，外观采用法国文艺复兴风格，立面横向分为三段：一、二层基座为斩假石仿石墙面，中段用清水红砖，贯通的阳台和女儿墙构成檐部。

当年这座形态独特的高档公寓吸引了很多外国"金领"入住，如西门子公司的雇员。1941 年太平洋战争爆发后，英美侨民纷纷回国或被日军关入集中营，公寓多半空置，万国储蓄会的业务也日趋清淡。1949 年后，著名影星赵丹和秦怡曾在此寓居。

2010 年世博会前，武康路作为试点进行风貌道路整治，武康大楼也得到保护性修缮。外立面除掉了脏乱的雨篷，空调架的颜色细心涂成灰红两色，分别与其所处外立面部位的材质颜色一致，大楼默默地变美了。

除了武康大楼，邬达克还为万国储蓄会设计了巨鹿路的 22 幢美式住宅。2017 年，因业主装修破坏严重而引起关注的巨鹿路 888 号就是 22 幢小楼之一。这是邬达克 1918 年到上海后设计的最早的作品之一。1920 年，他在寄回家的明信片上，用草图和照片详细介绍了这个项目。当时，这位在"魔都"打拼的年轻建筑师还在家信中吐槽了工作感受："虽然在上海我设计的建筑数量比想象的多，但却无法实现大学时代的梦想——成为一名真正的建筑师……现在虽然我动动铅笔就可以指挥千军万马，但却再也无法回到设计小教堂时那种内心的宁静……这里的建筑师的理念和故乡的砖瓦匠们不相上下。"

参观指南

大楼外观和门厅可供参观，室内不对外开放。建议探访巨鹿路 852 弄 1—10 号、巨鹿路 868—892 号的美式别墅。

The Normandie Apartments are perhaps the most famous buildings on Wukang Road. Signature works by architect Laszlo Hudec, the apartment building is often referred to as the "little Flatiron" in reference to the famous New York skyscraper which it resembles.

Hudec designed for two influential banking institutions in old Shanghai — the Joint Savings Society (JSS), owner of the Park Hotel, and the International Savings Society (ISS) that invested in building the Normandie Apartments.

The two masterpieces were both listed among the "99 reasons for loving Shanghai" in 2014 when Hudec was voted as a "Shanghai Symbol" by a million Shanghai netizens.

The International Savings Society was founded in Shanghai in 1912 by several French merchants including Jean Beudin, whose residence on Fenyang Road was also designed by Hudec.

According to the book *Men of Shanghai and North China* published in 1933, Paris-born Beudin arrived in Shanghai on September 6, 1908 and until 1912, when the International Savings Society was started, was a partner in the firm Cohen & J. Beudin. He had previously seen military service with the 16th Regiment of Colonial Infantry and was awarded the "Chevalier of the Legion of Honour."

With its hugely successful lottery-linked deposit account, bank accounts soared from 191 in 1914 to 131,800 in 1934. In 1934 the society's savings deposits accounted for more than one-fifth of all savings in China. They then invested the money in bonds, foreign currencies and real estate, including the Picardie Apartments (today's Hengshan Picardie Hotel), the Gascogne Apartments and the Normandie Apartments.

The Normandie Apartments are in a favorable location and the building stands out in the area. The triangle lot is located at the intersection of five streets and provides an open view. The steep, sharp-headed building suits the peculiar shape of the plot and its appearance is suggestive of a flat-iron or powerful warship.

The eight-story building has a French Renaissance style. The external walls on the ground and second floors are covered in artificial stones while the floors above feature red bricks. A stone, ashlar surface on the top floor corresponds with the base. The cantilevering balconies and the parapets form eaves with double parallel corners. The three sections of balconies form a vivid waistline on the facade.

A 1920s English newspaper describes the building: "(It) contains 60 apartments, some with only one room besides kitchen and bath and others with hall, two bedrooms, living rooms, dining room, kitchen, etc, more in keeping with a family apartment."

It also has three elevators and several fire staircases. The two elevators in the entrance hall still show the floor number with pointers.

The building was one of the city's earliest high-end apartments and was initially leased to senior employees of foreign companies such as Siemens.

Following World War II, the Kuomintang government purchased the International Savings Society's properties around 1946. Famous Chinese artists including actor Zhao Dan and actress Qing Yi also lived in the building at one time.

Route Ferguson was renamed Wukang Road in 1943 and the Normandie Apartments became known as the Wukang Building in 1953.

The building was renovated before 2010 along with Wukang Road. Shanghai-based Zhang Ming Architectural Design Firm, which had previously renovated Hudec's Grand Theatre and Moore Memorial Church, cleaned the façade, restored the brickwork and designed frames to conceal the exterior air conditioning units. The signature arcade and well-designed entrance hall have both been restored.

While designing the Normandie building and other projects for the International Savings Society, Hudec was sending detailed sketches and descriptions of the project to his father on the back of some hand-made postcards. He also complained about his early Shanghai career in family letters.

"I won't be a real architect in Shanghai as I had imagined at the Technical University, although I build more than I would ever have thought... Now, when with a move¬ment of the pencil I direct thousands, I don't feel the same inner peace as when I drew those little chapels... The professionals around here have a concep¬tion of architecture like the average mason at home," Hudec wrote.

The "flatiron" edifice looks like a delicate, sophisticated mini city with so many long, zigzagging passageways.

It is quiet but filled with traces of life. Each long passageway seems to be a gallery exhibiting a distinct Shanghai way of life. There are big potted green plants, motorbikes covered by worn raincoats, an antique Chinese table and a rainbow of clothes hanging on a bamboo pole. Hudec's choice of a beige-toned mezzanine floor has endured and fits attractively in its surrounding.

At dusk on a spring day, a dash of yellow light leaks from one of the top-floor windows revealing an enduring warmth and depth to this old "magical city".

More than 90 years ago, a homesick Hudec drew blueprints for this elegant building. Did he know that it was destined to one day become a symbol of Shanghai?

Tips

The appearance and the entrance hall can be appreciated. The interiors of the suites are not open to the public. I would suggest visiting the lesser-known, very tranquil 22 American-style villas, which are now No. 1 to 10 on 852 Julu Road and No. 868 to 892 Julu Road. It will be a lovely walk to find them.

山林城市

与武康路有点缘分。

2009年我搬到武康路之后,偶然从母校南京金陵中学校史中得知,武康路筑路人、美国传教士福开森居然是母校的创始人。1888年,福开森受美国基督教美以美会邀请,在南京创办汇文书院,是金陵中学的前身。

我在他亲手设计的钟楼里上过实验课,那是南京一座美丽的19世纪建筑,同一间教室里曾经坐着拖着长辫子的晚清学生。

1896年,福开森离开南京到上海,出任南洋公学监院,参与创建工作。南洋公学是上海交通大学的前身,研究徐家汇时我又一次在档案里看到了福开森——西装革履的他与穿马褂的中国师生们合影。

可能是为了让这些寓居公共租界的同事们上班方便些,他修筑了福开森路,武康路的故事就是这样开始的。

在福开森告别金陵钟楼到上海的19世纪90年代,另一位来华的美国传教士佛礼甲正从杭州沿运河寻求避暑之地。他辗转来到武康(现为德清县)莫干山麓,被翠竹、泉瀑和山居的幽静深深吸引。

几位英美传教士很快接踵而至,筑舍避暑,历史悠久的莫干山作为"新避暑地"被广为报道,成为那个时代的"网红"。1937年版的《莫干山导游》提到,佛礼甲发现莫干山时还是座荒山,攀登不易,"他未预料有今日之盛"。

1930年的《字林报行名簿》显示,武康路115号密丹公寓的几位洋居民供职于石油和建筑等行业,都在外滩一带上班。进一步研究老上海英文报纸,我发现在外滩工作,在武康路一带居住,到武康莫干山度假,曾是"魔都"金领让人羡慕的生活方式。

为了弄清楚武康路是否真与浙江武康的莫干山有相似之处,2017年4

月，我专程途经武康县到莫干山度了一个周末。在山居民宿夯土小屋的露台上，我泡了一杯清茶，眼前满是绿树翠竹，看不到人，只有风吹竹叶的声音。自然美的意境与武康路真有异曲同工之妙。

除了环境和氛围，武康路和武康莫干山的发展轨迹也是神一样的同步。在黄金的 20 世纪 20 年代，福开森路与莫干山麓的建设活动同样繁忙，都是先由外国侨民建起一座座风格各异的西式住宅，然后渐渐地群贤毕至。

2007 年武康路开始实验性风貌保护与改造时，南非人高天成骑车来到莫干山，也被翠竹山林深深地打动。为了享受回归山林的时光，他在莫干山竹林中建了朴素的农舍，后来发展为著名的"洋家乐"——"裸心乡"和"裸心谷"，推动了莫干山民宿热的兴起。莫干山和武康路再次复兴的背后，不正是忙碌都市人对既有现代化舒适体验，又不失自然美的山林城市的追求吗？

福开森先生肯定也未预料到，这条曾以他命名的路会有"今日之盛"。

乔争月
2017 年 11 月 于武康路月亮书房

A Garden City

After I moved to Wukang Road in 2009, I happened to discover that John Calvin Ferguson, for whom the former French Concession's Municipal Administrative Council named the road Rue de Ferguson, was also the founder of my middle school. Commissioned by Methodist Episcopal Church, this American missionary opened a school in my hometown Nanjing which is today's Nanjing Jinling Middle School where I studied six years.

I took biology classes in the old bell tower, a beautiful 19th-century building designed by Ferguson himself. Archive photos show the same classroom filled with boy students, all with their long pigtails, sitting for class during the Qing dynasty.

In 1896 Ferguson left Nanjing for Shanghai on an invitation to found the Nanyang College, which has since developed into the Shanghai Jiao Tong University. When researching and writing about Xujiahui, I did find him in the archives — a group photo of Ferguson attired in a western suit with Chinese faculty and students in mandarin jackets.

Probably to make it convenient for his fellow colleagues to travel from their homes in the downtown international settlement, he helped fund the construction of a small road which later became the Route de Ferguson. This is where the story of Wukang Road began.

In the 1890s, as Ferguson bid farewell to the bell tower for Shanghai, another American missionary, presumably Dr. Farnham, was traveling from Hangzhou along the river in search of an ideal summer resort. The missionary's Chinese name was "Fo Li Jia", and the academic discussion over his identity persists even now. However when he came to the foot of Mount Moganshan in the then Wukang County, he was enamored by the green bamboo, springs, falls, and overall tranquility of the nature there.

Several American and British missionaries arrived soon thereafter and built summer houses, giving Mount Moganshan a reputation as the "new summer resort." According to the 1937 version of *Moganshan Guide*, Mount Moganshan was still a deserted mountain and challenging to climb when missionary "Fo Li Jia" discovered it. The 1937 book notes that "he hadn't expected the mountain would enjoy the prosperity today."

According to the *Shanghai Directory* published by the *North China Daily News* in 1930, foreign residents of the Midget Apartments on 115 Wukang Road worked in petroleum or construction firms, which were all on or near the bund. Further research among old Shanghai English newspapers showed some middle or upper class local expatriates had enjoyed an attractive work-life pattern. They worked in the bund area, lived in the Wukang Road neighborhood and went to Wukang County to spend holidays on the Moganshan Mountain.

In the spring of 2017, I made a special trip traveling through the former Wukang County for a weekend in Mount Moganshan, just like some of my neighbors in old Shanghai would do some 80 years ago. On the balcony of a hill hut, I steeped a cup of green tea to enjoy a moment of tranquility. Fronted with a forest of emerald bamboos, I could hear the wind breezing through the leaves. The natural beauty is reminiscent of the style of Wukang Road itself.

In addition to similar environment and atmosphere, Wukang Road and Mount Moganshan seemed to experience a parallel evolution. In the 1920s both Route de Ferguson and Moganshan witnessed much construction, where some expatriates first came to build western-style houses and gradually more people were attracted to come and build more.

When Wukang Road kicked off the experimental regeneration in 2007, Grant Horsfieldn, a South African, bicycled to Moganshan. He was also deeply attracted by green bamboo and tranquil forests. To enjoy a life back to the nature, he developed pristine farmhouses to be the famous resort "Naked Retreats" which promoted a mushrooming of stylish homestays all over the mountain.

Whatever the reasons behind the thriving of Mount Moganshan and Wukang Road 80 years ago or today, in my mind it is all the same, the busy urban dwellers' pursuit for a garden city that well incorporates both natural beauty and modern comforts.

Perhaps Mr. Ferguson also hadn't expected the road named after him would enjoy the prosperity today.

Michelle Qiao
November 2017
Moon Atelier, Wukang Road

推荐阅读　Recommended Readings

一、专著

1 沙永杰 纪雁 钱宗灏 著. 上海武康路—风貌保护道路的历史研究与保护规划探索. 上海：同济大学出版社，2009.

2 白吉尔 著. 上海史—走向现代之路. 王菊 赵念国 译. 上海：上海社会科学院出版社，2014.

3 方世忠 主编. 海上遗珍武康路. 北京：中华书局，2017.

4 梅朋 傅立德 著. 上海法租界史. 倪静兰 译. 上海：上海社会科学院出版社，2007.

5 伍江 著. 上海百年建筑史 1840-1949（第二版）. 上海：同济大学出版社，2008.

6 郑时龄 著. 上海近代建筑风格. 上海：上海教育出版社，1999.

7 罗小未 主编. 上海建筑指南. 上海：上海人民美术出版社，1996.

8 陈从周 章明 著. 上海近代建筑史稿. 上海：上海三联书店，1988.

9 许乙弘 著. Art Deco 的源与流：中西摩登建筑关系研究. 南京：东南大学出版，2006.

10 蔡达峰 宋凡圣 主编. 上海近代建筑史稿. 南京：江苏文艺出版社，2013.

11 华霞虹 乔争月等. 上海邬达克建筑地图. 上海：同济大学出版社，2013.

12 卢卡·彭切里尼 邬达克. 华霞虹 乔争月 译. 上海：同济大学出版社，2013.

13 福开森 著. 中国艺术讲演录. 张郁乎 译. 北京：北京大学出版社，2015.

14 Tess Johnston & Deke Erh. *A Last look: Western Architecture in Old Shanghai*. Shanghai: Old China Hand Press,1993.

15 F. L. Hawks Pott D.D. *A Short History of Shanghai*. Beijing:China Intercontinental Press, 2008.

16 Harriet Sergeant. *Shanghai*. London: Jonathan Cape Ltd, 1990.

17 Kate Baker, Patrick Cranley, Spencer Dodington, Tess Johnston, Tina Kanagaratnam and Carolyn Robertson. *Final Five Shanghai Walks*. Shanghai:Old China Hand Press,2015.

二、近代英文报刊

《字林西报》 *North China Daily News*

《北华捷报》 *North-China Herald*

《大陆报》 *The China Press*

《密勒氏评论》 *The China Weekly Review*

图片来源　　Image Source

P4-5、P6-7、P8-9
Virtual Shanghai, http://www.virtualshanghai.net/Photos/Images

P32
艾潇.百年足迹——西安交通大学110年[M].西安：西安交通大学出版社，2006.

P52
《蔡声白先生传略》[出版情况不详]

P57
同济大学建筑与城市规划学院.谭垣纪念文集[M].北京：中国建筑工业出版社，2010.

P76、P77
郑时龄.上海近代建筑风格[M].上海：上海教育出版社，1999.

P84、P85
Collar,Huge,.Woodroffe,pauline. *Captive in Shanghai*. Hongkong:Oxford University Press,1990.

P96
Sowerby,Arthur de Carle. *China's Natural History: A Guide to the Shanghai Museum*. Shanghai:Royal Asiatic North China Branch,1936.

P97
Wilkinson,Edward Sheldon. *Shanghai Birds:a study of bird life in Shanghai and the surrounding districts*. Shanghai:North-China Daily News &Herald,1929.

P119
蔡达峰，宋凡圣，主编.上海近代建筑史稿[M].南京：江苏文艺出版社，2013.

P122
The China Press

P128
郑时龄.上海近代建筑风格[M].上海：上海教育出版社，1999.

P139
胡宗刚 著.北平研究院植物学研究所史略[M].上海：上海交通大学出版社出版，2011.

P145
宋庆龄纪念馆提供照片

P78、P79、P133、
乔争月 拍摄

其余图片由张雪飞、邵律 拍摄

图书在版编目（CIP）数据

上海武康路建筑地图 / 乔争月，张雪飞著. -- 上海：同济大学出版社，2018.8

（城市行走书系）

ISBN 978-7-5608-7972-7

Ⅰ.①上… Ⅱ.①乔… ②张… Ⅲ.①建筑物－介绍－上海 Ⅳ.① TU-862

中国版本图书馆 CIP 数据核字 (2018) 第 151947 号

上海武康路建筑地图

乔争月　张雪飞　著

出 品 人：华春荣
策划编辑：江岱
责任编辑：徐希
助理编辑：周希冉
责任校对：徐春莲
装帧设计：孙晓悦
出版发行：同济大学出版社 www.tongjipress.com.cn
地　　址：上海市四平路1239号 邮编：200092
电　　话：021-65985622
经　　销：全国新华书店
印　　刷：上海雅昌艺术印刷有限公司
开　　本：787mm×1 092mm　1/36
印　　张：4.5
印　　数：1－3 100
字　　数：134 000
版　　次：2018年8月第1版　2018年8月第1次印刷
书　　号：ISBN 978-7-5608-7972-7
定　　价：45.00元